THE CAC BOOMERANG

Australia's own WWII Fighter

DON WILLIAMS

Illustrations by Michael Claringbould

The CAC Boomerang
Australia's own WWII Fighter
Don Williams
Illustrations by Michael Claringbould

ISBN 9780975642320

First published 2024 by Avonmore Books

Avonmore Books
PO Box 217
Kent Town
South Australia 5071
Australia

Phone: (61 8) 8431 9780
avonmorebooks.com.au

 A catalogue record for this
book is available from the
National Library of Australia

Cover design & layout by Diane Bricknell

CONTENTS

ABBREVIATIONS & GLOSSARY

AFC	Australian Flying Corps
AWM	Australian War Memorial
BHAS	Broken Hill Associated Smelters
CAC	Commonwealth Aircraft Corporation
CAS	Chief of the Air Staff
DAP	Department of Aircraft Production
DFC	Distinguished Flying Cross
EFTS	Elementary Flying Training School
FCU	Fighter Control Unit
GMH	General Motors Holden
Kokutai	A Japanese naval air group
NAA	National Archives of Australia
ORB	Operational Record Book
OTU	Operational Training Unit
RAAF	Royal Australian Air Force
RAF	Royal Air Force
RNZAF	Royal New Zealand Air Force
RVAC	Royal Victorian Aero Club
Sentai	Japanese army flying regiment
TacR	Tactical Reconnaissance
US	United States
USAAF	United States Army Air Forces
WWI	World War One
WWII	World War Two

ABOUT THE AUTHOR

Don Williams

Don Williams grew up in the 1960s and 1970s, absorbing a host of television programs, books and movies about the titanic World War Two aerial battles in Europe's skies. His childhood bedroom was filled with models of Spitfires, Hurricanes, Messerschmidts and Stukas.

Later in life, Don sought to better understand the role of the Royal Australian Air Force in the Pacific War. He was surprised and disappointed to find that no book telling the complete story of Australia's own wartime emergency fighter, the Commonwealth Aircraft Corporation Boomerang, had been published.

Don decided to fill this gap by writing a book dedicated to the Boomerang. He is also a keen amateur photographer and is pleased that some of his photographs are included in the book.

Sound research and writing were foundations of Don's career as a policy analyst. These skills were further enhanced when he was awarded a PhD as a mature age student. He also researched and published an article that identified the Royal Air Force squadron responsible for the death of Nazi Germany's Field Marshal Fedor von Bock, the most senior German officer killed by enemy fire in World War Two.

ACKNOWLEDGEMENTS

The generous support, encouragement and tolerance of my wife Pamela were the *sine qua non* that made this book possible. I would also like to thank the family members and friends who bravely reviewed successive drafts of the manuscript and provided indispensable feedback.

Thanks also to Matt Denning, who restored Boomerang A46-122 to flying condition. Matt provided invaluable advice about the Boomerang's design and construction used in Chapter 6.

INTRODUCTION

This book tells the story of the Commonwealth Aircraft Corporation Boomerang, Australia's only locally designed and built fighter aircraft. The Boomerang has the further distinction of being the *sole* Australian-designed and produced aircraft to see combat, making it a vitally important part of Australia's military aviation history. It is thus surprising that no standalone publication dedicated to the Boomerang has been released, including aspects such as why it was needed, its development and production, operational use by the Royal Australian Air Force and accounts of pilots who fought and died flying it. This book is intended to fill that gap.

The book is based on contemporary first-hand records and the insights they provide. These records reveal the conflicting views put to Australia's War Cabinet in early 1942 about whether the Boomerang should be rushed into production to help Australia survive what was viewed as an existential crisis. The reasons why the War Cabinet subsequently approved additional orders for an aircraft no longer wanted by the RAAF are also examined.

The Boomerang's operational career is outlined, including its successful transformation from undistinguished interceptor fighter to successful army cooperation aircraft. The flexibility and skill of the RAAF pilots and groundcrew responsible for this transformation are highlighted, as are the dangers faced by Boomerang pilots.

By providing a clear understanding of the Boomerang's origins, production and operational career the book will draw a balanced assessment of the Boomerang's successes and failures.

I focus specifically on the Boomerang and do not examine the development of higher-powered, higher-performance derivatives of the original aircraft. These developments are of great technical interest but ultimately produced no aircraft that flew operationally in World War Two.

Interestingly, the rushed development of the Boomerang illustrates the difficulties and risks associated with procuring military equipment. Although the Boomerang saw active service in the middle of the previous century, it provides lessons in military equipment acquisition that are relevant in today's world of rising international tensions.

In telling the Boomerang's story I have depended largely on primary records held by the National Archives of Australia. These include documents from the Australian Government's War Cabinet, the Royal Australian Air Force and the Commonwealth Aircraft Corporation.

The Boomerang took to the air decades ago during WWII. None of the active participants in the story, whether politicians and industrialists at the highest decision-making levels or front-line pilots, are still alive. To bridge this gap and emphasise the human aspects of the Boomerang story, the book includes contemporary first-hand accounts, including reports from pilots returning from operational missions and casualty reports. We must remember that the pilots were young men, mostly in their twenties, who frequently lost their lives when aircraft failed to return. Placing this human cost on the record is the most important part of any book about military aviation, including this one.

Don Williams
Melbourne
March 2024

MAPS

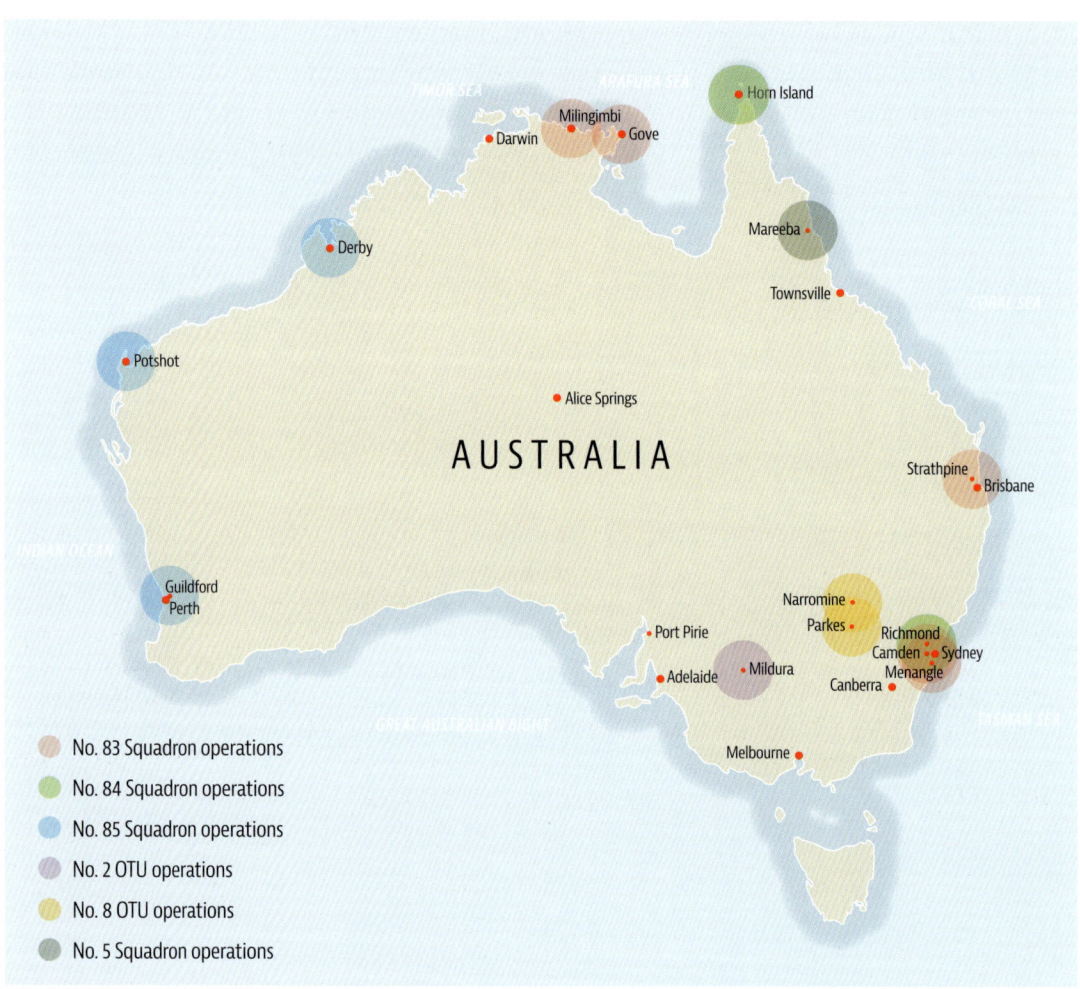

This map shows locations in Australia where Boomerangs operated. Two training units, Nos. 2 and 8 OTUs, flew Boomerangs from Mildura, Narromine and Parkes. In 1943 Nos. 83, 84 and 85 Squadrons were established at Strathpine, Richmond and Guildford respectively. These units subsequently operated the Boomerang in the Interceptor Fighter role from frontline locations in Western Australia, the Northern Territory and Horn Island in Far North Queensland. No. 84 Squadron attempted to intercept enemy aircraft from Horn Island and also from nearby Merauke in Dutch New Guinea (see the map on p. 9), while No. 85 Squadron attempted interceptions from Potshot. In late 1943 No. 5 Squadron received Boomerangs at Mareeba where they flew training missions in the army cooperation role for a year before deploying to Torokina in Bougainville (see p. 9).

This map shows the approximate location of Boomerang operations in the New Guinea area. As noted on the previous page, No. 84 Squadron briefly operated from Merauke in Dutch New Guinea. However, Boomerangs saw the vast majority of their frontline service in the army cooperation role with Nos. 4 and 5 Squadrons. Headquartered in Port Moresby, from July 1943 No. 4 Squadron began flying Boomerangs from forward airstrips in the areas of northern New Guinea shown. The Boomerangs were heavily utilised until late March 1945 when they deployed to Borneo where they supported Australian Army operations until the end of the war. Meanwhile, No. 5 Squadron deployed Boomerangs to Torokina in Bougainville in late 1944 where they operated until the end of the war. After No. 4 Squadron deployed to Borneo in 1945, No. 5 Squadron sent detachments to New Britain and New Guinea.

A fine aerial image of No. 5 Squadron's A46-146 over Mareeba in 1944. (AWM)

CHAPTER 1
STRATEGIC LEADERSHIP OF AUSTRALIA'S AIR WAR IN WWII

The Boomerang story was heavily influenced by the groups of high-ranking military and political leaders who directed Australia's air war in WWII. These groups and their responsibilities are summarised below:

Body	Responsibility
Australian War Cabinet	During WWII Australia's main decision-making body was the War Cabinet, established when the war started in September 1939. The War Cabinet was initially chaired by Prime Minister Robert Menzies (United Australia Party). When the Australian Labor Party, led by John Curtin, assumed power in October 1941 the War Cabinet continued to make the major decisions about the conduct of the war. It was the Curtin-led War Cabinet that made crucial decisions about the Boomerang. The War Cabinet only included members of the government. An Advisory War Council was established to ensure both government and opposition parties were involved in directing Australia's war effort.
Advisory War Council	The Advisory War Council also had an important decision-making role. The Council, chaired by the Prime Minister, included members of the government's War Cabinet, the Leader of the Opposition and three members of the Opposition. It allowed the government and the Opposition to work together on key decisions about the war. The Council was established after John Curtin (then leader of the Opposition) rejected a proposal to convene a national unity government in September 1940. From October 1941 Advisory War Council decisions were accepted as War Cabinet decisions, giving the Council tremendous authority. In the context of this book, decisions about Boomerang production were generally considered by the War Cabinet without necessarily being referred to the Advisory War Council. Specific decisions about the Boomerang that were considered by the Advisory War Council are discussed in the text.
Air Board	The Air Board was responsible for controlling and directing the RAAF from the Board's creation in 1921 until its abolition in 1976. The Board included both senior RAAF officers and civilian appointees. The Board's composition changed regularly and by mid-1942 it included the Chief of the Air Staff, an Air Member for Personnel, an Air Member for Supply and Equipment, an Air Member for Engineering and Maintenance, a civilian Finance Member and a civilian Business Member.
Department of Aircraft Production	Recognising the crucial role of aircraft in the war, the government established a Department of Aircraft Production in June 1941. This led to the appointment of a Minister for Aircraft Production and a Director General of Aircraft Production. Donald Cameron was the Minister for Aircraft Production from October 1941 until February 1945. Cameron participated in key decisions about the Boomerang and we encounter him several times in this book. The very capable industrialist Essington Lewis was the Director General of Aircraft Production. We will also encounter him.

Body	Responsibility
Aircraft Advisory Committee	The Aircraft Advisory Committee was established in December 1941 to coordinate military aircraft construction and maintenance across the entire Australian aviation industry. Its membership included representatives from the RAAF, the Department of Aircraft Production, the aircraft industry, trade unions and the Treasury. The Aircraft Advisory Committee provided advice to the Director General of Aircraft Production, who in turn advised the Minister for Aircraft Production: **War Cabinet** **Minister for Aircraft Production** **Director-General of Aircraft Production** — **Aircraft Advisory Committee** **Department of Aircraft Production**

The inaugural meeting of the Australian War Cabinet on 27 September 1939. At the end of the table in the centre is Prime Minister Robert Menzies. (AWM)

CHAPTER 2
WHY AUSTRALIA NEEDED AN EMERGENCY FIGHTER IN WWII

The Commonwealth Aircraft Corporation Boomerang is the only military aircraft designed, produced and used on active service by Australia. The Boomerang was conceived as an "emergency fighter" in December 1941 when Japan launched a full-scale war of expansion in the Pacific and South East Asia. Facing an apparently existential crisis, Australia was close to defenceless against air attack. In a desperate bid to shore up Australia's security the Boomerang moved from the drawing board to first flight in an astonishingly short time, making its initial flight in May 1942. A brief review of the Royal Australian Air Force (RAAF) between the First and Second World Wars helps to explain why Australia desperately sought to develop an emergency fighter at this moment of national crisis.

Australia's population in the first half of the twentieth century was very small. It was just under 5,000,000 at the start of WWI, while Australia was located far from the world's great military powers. Despite these disadvantages the Australian Army's Central Flying School was founded in 1912 and an Aviation Instructional Staff was formed in 1913, which gave its first flying course at Point Cook in Victoria in 1914. The Australian Flying Corps, a branch of the Australian Imperial Force (Australia's WWI expeditionary force), fought with distinction on the Western Front and the Middle East in WWI. Australians also flew with the Royal Flying Corps, the Royal Naval Air Service and the Royal Air Force (RAF). The RAAF was established in 1921 and is by some accounts the world's second oldest air force. During the interwar period the RAAF's senior ranks included personnel who had experienced aerial combat first-hand and benefitted from its hard-won lessons.

One of these WWI veterans was Richard Williams, Chief of the Air Staff (CAS, then the RAAF's senior officer) in 1925. Williams identified the defeat of enemy aircraft as a key task for the RAAF. He recommended a force structure with a strong fighter component (124 aircraft from a total of 324) to gain air superiority. With remarkable foresight, Williams specifically suggested that Australia should consider how to defeat Japanese naval air power when it equipped its air force. However, the Australian government shelved Williams' proposals and commissioned the RAF's Air Marshal Sir John Salmond to review the RAAF in 1928.

Sir John's appointment typified Australia's status as a Dominion in the British Empire with a corresponding deference to Britain, not least in defence circles. Australia's armed forces essentially

The Air Board in 1928, with Air Commodore Richard Williams front centre. (AWM)

complemented those of Britain under the "Imperial Defence" doctrine that prevailed in the interwar period. In an abrupt change from Williams' thinking, Sir John dismissed the threat of major attacks on Australia by battleships or aircraft carriers, or large-scale landings. He considered that potential threats were limited to smaller-scale maritime "raids", which could be neutralised by bomb and torpedo attacks. A revised Salmond Plan was adopted by the government in 1936, which did include some recognition of the dedicated fighter role. This plan included three fighter or fighter-bomber squadrons for local air defence at key points on Australia's east coast in its proposed seventeen squadron order of battle.

In 1938, with international tensions rising, Australia again turned to Britain for advice about the RAAF's development. This came via the RAF Inspector General, Sir Edward Ellington, who had already provided advice about the RAAF's 1936 development program. A key element of Sir Edward's advice in both 1936 and 1938 was "ubiquity of purpose", which held that specialisation of aircraft should be avoided. Ubiquity of purpose would free the RAAF from the unwelcome prospect of being equipped with aircraft specifically designed to support ground and naval forces, which would have hampered its ability to carry out independent air operations.

In accordance with Sir Edward's recommendations the RAAF coastal reconnaissance squadrons became "general reconnaissance" units and local defence squadrons (which could fight enemy aircraft) became "general purpose" units, largely tasked with reconnaissance and strike duties rather than air defence. Sir Edward apparently intended to reinforce the RAAF's capacity to carry out independent air operations, but he did not highlight air defence and the associated need for high-performance fighters. This lack of focus on modern fighters would continue to corrode the RAAF's thinking after war broke out in 1939.

One constant in the RAAF's interwar existence was severely restricted funding, a problem that increased during the Great Depression in the 1930s. The RAAF also had to compete for scarce defence funding with the Navy and the Army. These financial restraints undoubtedly hindered the RAAF's efforts to achieve a modern, balanced force structure. However, everything changed when the world again entered an all-consuming global conflict.

With the start of WWII in September 1939, the financial and resource constraints the RAAF had suffered were suddenly overturned. The RAAF established a Home Defence Force, which CAS Air Vice Marshal James Goble recommended be brought to a strength of nineteen squadrons by June 1940. These nineteen squadrons consisted of:

- Six general reconnaissance squadrons
- Six general purpose squadrons
- Three army cooperation squadrons
- One fighter squadron
- One fleet cooperation squadron
- Two flying boat squadrons

In April 1940 the War Cabinet considered a recommendation by Air Minister James Fairbairn to re-allocate the sole proposed fighter squadron to the general reconnaissance role, using Lockheed Hudson maritime patrol bombers. Fairbairn's reasoning, endorsed by the chiefs of Australia's armed forces, relied on the "continuous lessening of the probability of attack on Australian territory by Japan" so that "the possibility of attack by carrier borne aircraft against this country is remote". Under these conditions the primary RAAF Home Defence tasks would be locating and destroying hostile ships, and convoy protection in conjunction with the Navy. The only aircraft likely to be launched by hostile raiding ships would be reconnaissance types, which would not be able to defeat attacking RAAF bombers. This meant that RAAF aircraft attacking hostile ships would not need fighter escorts. The War Cabinet's approval of Fairbairn's recommendation left the RAAF's Home Defence force without a single fighter squadron.

Another notable insight from Fairbairn's submission to the War Cabinet is that it had been intended to equip the single Home Defence fighter squadron with the unwieldy twin-engine Bristol Beaufighter instead of a single-engine single-seat fighter, which was the state-of-the-art fighter type at the time.

In June 1940 the War Cabinet considered RAAF recommendations to further expand the Home Defence Force. An additional thirteen squadrons were proposed, which would bring the Home Defence Force to an impressive 32 squadrons. These thirteen additional squadrons would include seven general reconnaissance squadrons, three general purpose squadrons, one flying boat squadron and two fighter squadrons. The War Cabinet endorsed the recommendations. Although the 32-squadron force included a dedicated fighter component, the associated acquisition program assigned the highest priority to bombers and the lowest to fighters.

Air Minister James Fairbairn who helped shape the initial wartime structure of the RAAF. He was killed in an air crash in August 1940.

Aircraft priority for the 32 squadron Home Defence Force, June 1940 (highest priority listed first):

- 71 Hudsons (or equivalent) for existing General Reconnaissance squadrons in Australia

- 162 General Purpose two-seat dive-bombers to re-arm existing General Purpose squadrons

- 27 Amphibians for Fleet Cooperation squadrons

- 75 Hudsons (or equivalent) for new General Reconnaissance squadrons

- 81 General Purpose two-seat dive-bombers for new squadrons

- 54 long range two-seat fighters for new squadrons

The fighter squadrons were again allocated "long range two-seater fighters" (effectively Beaufighters) instead of high-performance single-seat fighters.

During 1941, despite increasing tensions between Western powers and Japan, the conviction that Australia's defence should focus on countering minor seaborne raids remained unchallenged. In a further display of deference to Britain, the RAAF's January 1941 Air War Effort (the overall plan for the organisation and strength of the RAAF) was submitted to the British Air Ministry for review. The Australian Air Staff's June 1941 response to the (generally favourable) Air Ministry comments identified the main tasks of the RAAF as maritime reconnaissance, attacks on shipping targets, support of ground forces and embarking aircraft on the Navy's ships. Providing the capacity to achieve air superiority by high-performance fighter aircraft was conspicuously absent from this list.

In the early 1940s air superiority was won by single-engine, single-seat fighters. Setting aside the deceptively glamorous aura associated with fighters, their principal purpose in aerial warfare was to engage and destroy enemy aircraft. This allows air superiority or, ideally, air supremacy to be gained.

Large-scale air combat in the early years of WWII demonstrated the unparalleled effectiveness of single-engine, single-seat fighters, such as Britain's Supermarine Spitfire and Hawker Hurricane, Germany's Messerschmidt Bf 109 and Japan's Mitsubishi A6M Zero. These aircraft were flown by a single pilot and fitted with either machine guns or a combination of machine guns and larger calibre cannons. They were fast by the standards of the day, had good acceleration, were manoeuvrable and could climb rapidly to engage attacking aircraft as quickly as possible.

The Battle of Britain in 1940 was won by the RAF's highly capable Spitfire and Hurricane fighters, flown by highly skilled pilots. The fighters were controlled by an advanced command and control system based on the new tool of radar. The RAF's cutting-edge air defence system disproved a well-known 1930s saying that "the bomber will always get through".

Other fighter configurations such as larger twin-engine, two-seat aircraft lacked satisfactory all-round performance. A famous example of this type, Germany's Messerschmidt Bf 110, was outfought by the RAF's Spitfires and Hurricanes in the Battle of Britain. The larger Bf 110 lacked the agility and acceleration of the smaller RAF fighters and generally could not compete with them on equal terms in air-to-air combat.

By the start of the Pacific War in December 1941 it should have obvious that failing to equip the RAAF with single-engine, single-seat fighters would leave Australia terribly vulnerable to air attack.

CAS Richard Williams had put the case for a strong fighter component for the RAAF as far back as the mid-1920s. Unfortunately, this far-sighted recommendation was ignored. Prior to WWII, the RAAF, RAF advisers and successive Australian governments were instead gripped by two fatally flawed assumptions. A key error was the assumption that attacks on Australia would be confined to maritime raids of limited scale. The possibility of large-scale attacks by aircraft carriers was dismissed as remote. This flawed assessment led to the limited resources available to the RAAF being focused on acquiring aircraft to locate and attack raiding vessels with bombs and torpedoes. Remarkably, the threat of attack on Australia by Japan was assessed as reducing in 1940. This misconception led to a belief that the need for high-performance fighters was minimal. A further error was to discount the possibility of attack by land-based aircraft. The idea that aircraft could attack Australia from bases within a thousand kilometres of the mainland, as took place in early 1942, was simply unimaginable.

These deeply flawed assumptions by Australia's defence establishment led to the RAAF's Home Defence Force being without any front-line fighter aircraft at the start of the Pacific War. The need for an emergency program to acquire single-engine, single-seat fighters was born.

In this moment of crisis, a mechanism to provide home-defence fighters to the RAAF within months was put to the Australian government by the Commonwealth Aircraft Corporation, the dominant company in the local aircraft industry.

Air Marshal Richard Williams, right, in London in 1941 when serving as Air Officer Commanding, RAAF Overseas Headquarters. As CAS in the late 1930s Williams had been instrumental in facilitating Wirraway production. (AWM)

A wartime photograph of industrialist Essington Lewis.

CHAPTER 3
A BATTLE WON: APPROVAL TO PRODUCE AN AUSTRALIAN INTERCEPTOR FIGHTER

There would have been no Australian WWII fighter without a local military aviation industry. It's perhaps surprising that a nation with a small population (around 7,000,000 in 1940) and narrow industrial base was able to design and build the Boomerang in the early 1940s. Events in the 1930s explain why this was possible.

In the years after its formation, the RAAF was equipped with aircraft imported from Britain. However, during the 1930s, astute observers grasped the unpleasant reality that this supply could be disrupted by increasing tensions in Europe and the urgent need to strengthen Britain's RAF. By October 1935 Minister for Defence Archibald Parkhill was warning his colleagues that aircraft were not being obtained from Britain within a reasonable time. He was concerned that "under certain circumstances it is not likely that Australia could hope to obtain any aircraft at all from Great Britain". This grave threat to the RAAF's future could be overcome by establishing a local aircraft manufacturing capacity. CAS Richard Williams confirmed the excessive delays in receiving aircraft from Britain. Parkhill was asked in Parliament in March 1936 about what the government was doing to encourage the manufacture of military aircraft in Australia. In the same month the *Sydney Morning Herald*, Australia's oldest newspaper, captured the mood when it warned that:

> ... fighting machines ordered from England will not be delivered for eighteen months at the earliest ... the making of fighting machines in Australia must now be regarded as an emergency and critical obligation.

Surprisingly, with threats to the umbilical cord tying the RAAF to Britain becoming more visible, the initial impetus for the creation of an Australian military aircraft industry came from the private sector instead of government. In 1934 Australia's leading industrialist, Essington Lewis, toured Europe, the United States and Japan. He was gravely concerned about Japan's growing industrial strength and aggressive outlook. Lewis also believed that airpower would play a crucial role in any new conflict. Lewis shared his ideas with other leading industrial figures on his return home including William Robinson, the managing director of mining and smelting firm Broken Hill Associated Smelters (BHAS) and Laurence Hartnett, the managing director of vehicle manufacturer General Motors Holden (GMH). Robinson and Hartnett sympathised with Lewis's views.

Lewis also voiced his concerns to government. In April 1935 Defence Minister Parkhill:

> ... made a tour of the Newcastle group of various factories ... talked with Essington Lewis ... he was very disturbed about the defences of Newcastle – the heart of Australian industry – not a defence gun, aeroplane or other weapon in sight or promised, and the Newcastle community getting very alarmed. For his part, he wanted to get on and do something about it.

Reflecting the concerns raised by Lewis, Robinson, Hartnett and staff from BHAS and GMH discussed the local manufacture of military aircraft with senior government officials in August 1935. The government representatives included CAS Williams. This meeting followed previous informal discussions between the companies and government officials on the topic. Clearly these discussions were productive, because Parkhill wrote to the companies in October 1935, inviting them to submit proposals to government for the establishment of a local industry to manufacture aircraft and engines. GMH's Hartnett replied, stating that the companies had established a syndicate to advance the manufacturing proposal and were asking Parkhill for support. The path to establishing a local military aircraft industry now took a detour while the Australian and British governments grappled with the inclusion of a "foreign" (i.e. the American dominated GMH) company in the syndicate. This issue was resolved by including additional British/Australian companies in the syndicate to dilute GMH's holding. This allowed Parkhill to write to the syndicate in June 1936, promising that:

> ... the Government will afford such assistance as lies within its power in fostering such an enterprise.

Events moved quickly after this endorsement. The companies in the syndicate advanced funds to establish the Commonwealth Aircraft Corporation Pty Ltd (CAC), which was registered in October 1936. Construction of CAC's factory at Fishermans Bend in Melbourne commenced in 1937. Extensive construction of facilities to manufacture airframes and engines continued until 1943.

Even before the syndicate morphed into CAC, it appointed a mission that toured aircraft manufacturers in Britain, the United States, France, Holland, Germany and Czechoslovakia in 1936. The mission's brief was to recommend an aircraft and engine that would provide useful assets for the RAAF, while reflecting Australia's level of industrial development. The aircraft/engine combination should also allow modern, economical production techniques to be introduced. The mission consisted of the prominent Australian aviation figure Lawrence Wackett and RAAF engineering officers Herbert Harrison and Arthur Murphy.

The mission recommended production of a low-wing monoplane with an enclosed cockpit, stressed metal mainplane and variable pitch propeller powered by a radial engine. To meet these criteria the mission further recommended licence production of North American Aviation's two-seat, fixed-undercarriage

The new CAC factory in March 1938. It was located at Fishermans Bend on the Yarra River, just a few miles from central Melbourne. The engine plant is on the left while the aircraft factory is on the right. (State Library of Victoria)

NA-16 advanced trainer. The mission also proposed that the aircraft be powered by a Pratt and Whitney single-row radial engine.

Wackett conveyed the mission's recommendations to CAS Williams, who endorsed them. Defence Minister Parkhill took the proposal to Cabinet with Williams' support. The Cabinet overcame its

Wirraway A20-2 which was actually a NA-33 pattern aircraft assembled by CAC in 1938. (AWM)

ingrained aversion to ordering American aircraft instead of British types and agreed to order 40 NA-16 aircraft (subsequently amended to the retractable undercarriage NA-33 version) and 50 engines. These aircraft and engines were ordered from CAC in January 1937. An important step was the first test run of a CAC manufactured single-row Wasp radial engine, which took place on 21 December 1938.

North American's licence for the NA-16/33 permitted some changes to cater for Australian requirements. The CAC-produced aircraft included modifications to allow the installation of additional machine guns and to facilitate dive-bombing. The resulting aircraft was named the Wirraway (an indigenous Australian term meaning "challenge"). The first CAC-produced aircraft was flown on 27 March 1939 and deliveries to the RAAF commenced in July 1939. The RAAF issued the Wirraway to its general purpose squadrons.

Unfortunately, what can now be seen as a crucial missed opportunity coincided with the successful start of the Wirraway manufacturing program. During an assessment of twin-seat fighters for the RAAF in June 1938, CAS Williams identified the possibility of local production of a Wirraway derivative powered by the more powerful Pratt and Whitney Twin Wasp engine. This would provide an alternative to the range of British fighter aircraft being considered. Wackett (now CAC general manager) stressed the feasibility of CAC producing an aircraft of this type to the Air Board. By early 1939 CAC had sent specifications for a Wirraway fighter to Williams and suggested it could readily produce two prototypes if an order were placed. The prospect of timely production of a locally designed fighter was agonisingly close to reality when Williams recommended ordering these prototypes to the Air Board in February 1939. However, consideration of the recommendation was deferred. The reasons for this deferral were made crystal-clear by the Air Board in May, when it explicitly recommended against proceeding with the Twin-Wasp powered Wirraway fighter. The Board considered this aircraft would be unsuitable for the long-range, over-water missions needed to counter the maritime raids that were the only perceived threat to Australia's security at the time. The limited performance of a Wirraway fighter was also mentioned. The Air Board chose not to pursue local production of fighters on these grounds.

A busy production scene at the CAC factory, with Wackett Trainer fuselages in the foreground and Wirraway fuselages behind them. (AWM)

Despite this decision orders for the Wirraway increased with the prospect and then the reality of war. A total of 491 had been delivered to the RAAF by December 1941. Production of a training aircraft designed by Wackett also commenced in March 1941 (known simply as the Wackett Trainer). As a result, CAC was firmly established as a military aircraft manufacturer by the time Japan brought war to Australia's doorstep in early 1942.

The appointment of Lawrence Wackett to lead the mission that identified an aircraft and engine suitable for production by CAC reflected his status as a dominant figure in Australian military aviation.

Lawrence James Wackett was born in Townsville in 1896. After completing his secondary education, he enrolled for officer training at the Royal Military College at Duntroon in Canberra. After graduating from Duntroon, he transferred to the Australian Flying Corps (AFC) and qualified as a pilot at the Central Flying School at Point Cook in Victoria in October 1915. Wackett was posted to No. 1 Squadron, AFC, which commenced operations in Egypt in April 1916. The squadron's duties included reconnaissance, bombing, strafing, photography and air to air combat.

Wackett's mechanical aptitude and initiative were quickly noticed by his superiors. In January 1917, he was given the responsibility of improving procedures to repair damaged aircraft and returning them to service more quickly. He carried out this role so efficiently that he was Mentioned in Despatches and subsequently was posted to the Royal Flying Corps' Orfordness Experimental Station in Britain. Wackett returned to operations with No. 3 Squadron, AFC, in France in 1918. In between flying reconnaissance missions, Wackett perfected procedures to supply ammunition to ground forces by parachute, an innovation used successfully in the Battle of Hamel. His outstanding service was recognised by the award of the Distinguished Flying Cross and the Air Force Cross.

Given Wackett's outstanding service record, mechanical knowledge and innovative mind, it was no surprise he was one of 21 officers assigned to the RAAF on its formation in 1921. Wackett qualified as an aeronautical engineer in 1923 and was appointed to head the RAAF Experimental Station on its establishment at Sydney in 1924. However, the station was closed in 1929 and a disgruntled Wackett resigned from the RAAF. He led the aircraft section of the Cockatoo Island naval dockyard from 1930 to 1934, then managed a small private aviation company (Tugan Aircraft) at Mascot in Sydney. By this time according to CAS Williams:

> Here is a man who could take a clean sheet of paper, design an aircraft, supervise its construction and test it in the air.

As well as appointing Wackett to lead the 1936 mission, the aircraft manufacturing syndicate bought out Tugan Aircraft to secure the services of Wackett and his key colleagues. When CAC was established in October 1936, Wackett was appointed as its inaugural managing director. Wackett wasted no time bringing his assertive, self-confident attitude to bear: the contract for the purchase of the land for the Fishermans Bend factory from the Victorian government was finalised in May 1937, but Wackett had already commenced construction of the plant a month beforehand.

CAC had successfully delivered hundreds of Wirraways and aircraft engines by the start of 1942, and manufacture of the Wackett Trainer had commenced in early 1941. It was all led by the experienced, formidable Wackett. No other organisation with a remotely similar capacity to design and produce fighter aircraft existed in Australia.

Laurence Wackett as CAC general manager.

CAC's unique capability is confirmed by looking at the Australian aviation industry as a whole. The government had foolishly accepted the March 1939 recommendation of a British Air Mission to establish a new organisation based on state railway workshops to produce the Bristol Beaufort light bomber for the RAAF and RAF. Absurdly, with CAC's factory underutilised in the absence of fresh orders, management and workers with no aviation industry experience were co-opted to an entirely new enterprise to carry out Australia's most technically advanced manufacturing project to date. The resulting Beaufort Division of the Department of Aircraft Production (DAP) was totally consumed with producing the bomber. Against all odds, the first flight of an Australian-produced Beaufort took place in August 1941.

De Havilland Australia, the local arm of Britain's famous de Havilland Aircraft Company, was a smaller operation based at Bankstown in Sydney. It was occupied with producing the company's renowned Tiger Moth trainer for RAAF service and for export (deliveries commenced in May 1940). Beyond CAC, the Beaufort Division of the DAP and de Havilland, no other aircraft manufacturing industry existed in Australia. When war arrived at Australia's doorstep the young nation's only hope of local production of fighter aircraft lay with CAC and its driven, ambitious managing director.

CAC's existence at the start of WWII can be attributed to conscious decisions taken by government and industry during the 1930s, followed by all the hard work and funding needed to establish any new large-scale industrial operation. In contrast, the presence of Boomerang chief designer Fred David at CAC was the culmination of an improbable chain of events that reflect the role of chance in all human endeavours, including war.

The uniquely credentialled Friedrich Wilhelm Dawid (Fred David) was born in Vienna on 17 February 1900 to Czechoslovak parents. After military service in WWI, David resumed studies and was awarded an engineering diploma in Austria in 1922 before working in Sweden in 1923–24.

As a Jew, David's next move had a double-edged outlook: it took him to the epicentre of advanced aircraft design in the 1920s and early 1930s but placed him in a Germany where the Nazi party's inexorable rise to power was well underway. He worked for the German aircraft manufacturer Heinkel from 1927 to 1933, apart from an interlude in the United States during 1929–30. At Heinkel David initially using his brilliant mathematical skills to carry out structural stress analysis. He subsequently joined the design team for the sleek, all-metal He 70 Blitz mail plane, which set several world speed records. The He 70's legacy carried into WWII: it influenced the design of the later He 111 medium bomber, which featured the elliptical wings and streamlined fuselage of the earlier aircraft. The He 111 was a mainstay of Nazi Germany's bomber force and also operated as a transport later in the war.

David married his German wife Else in Berlin on 10 May 1933. At other times the talented young aircraft designer and his wife could have looked forward to a future enlivened by personal and professional success. However, 1933 was no ordinary year in Germany. Nazi party leader Adolf Hitler was appointed Chancellor on 30 January 1933. The passage of the Enabling Act in March 1933 established Hitler as effective dictator, whose word was law. The relentless exclusion of Jews from all facets of public and professional life in Germany commenced in 1933. Suddenly David and Else's future was terrifyingly uncertain.

Fortunately, David had become a close friend of Ernst Heinkel (head of the Heinkel firm), who had been selling aviation technology to the Japanese aviation industry. Heinkel arranged for David to work as a consulting aerodynamicist at Japan's Aichi Watch and Electric Manufacturing Company, which had started producing aircraft in 1920.

At Aichi he worked on the design of the D3A Val dive-bomber. It may be no coincidence that the Val had the same elegant elliptical wings as the He 70. The Val saw extensive service as a dive-bomber with the Imperial Japanese Navy in WWII. It achieved notoriety during the attack on Pearl Harbor in December 1941 and the subsequent bombing of Darwin, amid many other operations during the first months of the Pacific War.

Although David had escaped the imminent threat of physical annihilation, he was a virtual outcast at Aichi, being allocated an isolated house and an office excluded from the main factory. David was bound to be receptive when Australian diplomatic sources contacted him and opened lines of communication with Wackett. The prospect of moving to Australia was persuasive. After leaving Japan, David arrived in Melbourne in March 1939.

Wirraway A20-21, an early production aircraft seen in RAAF service in February 1940. By this time Fred David was working as a CAC design engineer. (AWM)

David wasted no time in joining CAC, starting his employment with the company on 17 April 1939. In common with many newcomers to Australia he understood the benefits of adopting a more "Australian" name in what was then a very Anglo-Saxon culture, formally becoming Frederick William David in August 1939. He was working as a design engineer by 1940.

Fred David's journey from Nazi Germany to Japan and ultimately to Australia was truly remarkable. Equally remarkably, CAC gained the services of a talented and experienced aircraft designer, familiar with contemporary military aviation technology, just when these skills were sorely needed. However, an apparently existential crisis would soon test the limits of even David's capabilities.

A Japanese aircraft carrier task force attacked the US naval base at Pearl Harbor on 7 December 1941. Pearl Harbor was in fact one of a whole series of Japanese attacks, which included strikes on the British colonies of Malaya, Singapore and Hong Kong, and the US territories of the Philippines, Guam and Wake Island. These attacks were all carried out within seven hours. The United States, Great Britain and Australia declared war on Japan on 8 December.

The striking power and high performance demonstrated by Japanese carrier-based aircraft at Pearl Harbor ruthlessly exposed Australia's need for modern fighter aircraft. In this time of crisis, a ready-made solution appeared to be available. Fred David drew initial sketches of a fighter combining the Pratt & Whitney fourteen-cylinder Twin Wasp radial with a modified Wirraway airframe just three days after the attack on Pearl Harbor. Unlike CAC's earlier approach to matching the Twin Wasp with the Wirraway (a two-seater with a pilot and observer/gunner), David sketched a single-seat fighter, the configuration that proved decisive in winning air superiority in WWII. CAC authorised detailed design of the new fighter on 21 December 1941. Having made this decision, CAC established a design team overseen by chief engineer Bill Air for the new project. The team included David as chief design engineer, Alan Bolton (overall design and stressing), Joe Solvey (wing aerodynamics), Col Belwood (fan, spinner and cowling) and Ian Flemming (flight testing and aerodynamics). Morrie Lodge would be the project engineer responsible for manufacturing control.

The team had defined key aspects of the design by 24 December, including:

- maximise use of Wirraway components and production equipment
- the engine would be the Twin Wasp radial
- a target of 100 days from starting manufacture of the first aircraft to first flight
- no design changes during the first production run.

David completed a set of "Interceptor Project" specifications based on the Twin Wasp/ modified Wirraway airframe combination within nine days of CAC authorising the project. The resulting fighter would be armed with eight Vickers 0.303-inch calibre (7.7mm) machine guns with 700 rounds per gun.

The project was then elevated from the design office to the political realm of high-level defence policy. On 31 December 1941 Wackett wrote to Essington Lewis, recently appointed as Director General of Aircraft Production, vigorously promoting a scheme for rapid production of the aircraft. Wackett made typically bold claims about the merits of the project. He promised that, with sufficient engines and propellers being available:

… there will be no difficulty in producing 200 single-seat fighters, with first deliveries commencing in three months' time, the whole 200 to be delivered during the six months following the delivery of the first.

Wackett stated that the fighter could successfully counter carrier-based aircraft and would be very effective in the defence of cities and bases.

As Wackett intended, his proposal quickly gained interest at the highest levels of the Australian government. It was discussed by the Advisory War Council on 12 January 1942, resulting in Prime Minister John Curtin seeking additional information about the scheme. The RAAF's response to this request confirmed the desirability of producing an effective interceptor fighter but warned that production could be constrained by competing demands for the Twin Wasp engine.

The feasibility of a Wirraway-derived fighter was just one of many aircraft procurement decisions that confronted Australia in January 1942. Key decision makers from the Department of Air, Department of Aircraft Production and CAC met

Final assembly of a Wirraway at the CAC factory in May 1941. David's single-seat fighter design envisaged using as many existing Wirraway parts as possible.

The CA-4 bomber prototype which first flew in September 1941. The project ran into many technical difficulties and ultimately never entered production. However, it understandably consumed much time and energy at CAC when the "Wirraway Fighter" design was conceived. (AWM)

on 21 January to plan Australia's aircraft production policy. CAC was represented by Wackett and the company's chairman, industrial titan Harold Darling. The participants considered the availability of overseas aircraft, local production of airframes and local engine production.

The urgent need for fighters was raised by CAS Sir Charles Burnett. Sir Charles suggested that modern P-40 Kittyhawks could be sourced from the United States or local production could be considered. Essington Lewis stressed the value of local production as insurance against the possible unavailability of US fighters. Lewis also pointed out that local fighter production would provide war-related work for CAC's Fishermans Bend factory. In the absence of new orders, the factory would be underutilised from the cessation of Wirraway production in April until construction of the new CA-4 bomber started at the end of 1942.

CAS Burnett countered Lewis by noting that local fighter production could be constrained by the limited availability of engines. He also defended his territory by stressing the need for the Air Staff to assess the performance of the proposed "Wirraway-Interceptor". Given these concerns, he wanted the availability of fighters from the US to be clarified before recommending local production.

The meeting ultimately decided to support the production of 105 Wirraway-Interceptors, subject to engines being available. These would equip five squadrons under the RAAF's establishment tables at that time. The record of the meeting includes the following telling caveat:

> This decision was made after a very careful examination of all the risks associated with the ordering of an aircraft which had not yet been built and therefore had not been tested for performance by the Air Staff.

Behind the typically opaque bureaucratic language, this is a clear acknowledgement of

Burnett's concern about ordering local production of a fighter with unproven performance and his preference to exhaust other possibilities before proceeding down this path.

Despite this reservation, the proposal to proceed with the groundbreaking design and production of an Australian fighter aircraft was included in the program recommended by the meeting. The final step in realising an Australian fighter plane would be War Cabinet approval for production to commence.

A recommendation to produce 105 "Wirraway Fighter Interceptor" aircraft and engine and airframe spares was put to the Australian War Cabinet on 2 February 1942. Minister for Air Arthur Drakeford and Minister for Aircraft Production Donald Cameron made the recommendation as part of the broader aircraft production plan agreed at the 21 January meeting. As suggested at the meeting, the War Cabinet submission argued the CAC fighter would provide an insurance policy in case the urgent procurement of modern fighters from the US or Britain (also recommended to the War Cabinet) did not succeed.

But, in what must have been a very unwelcome decision for CAC, the War Cabinet did not immediately accept the recommendation that Wirraway Interceptors be produced. Prime Minister Curtin noted advice received on 31 January that 250 P-40 Kittyhawk fighters would be delivered to Australia from the United States. In view of this development and the potential diversion of resources from the production of other aircraft, the War Cabinet sought further information before making a decision about production of the Wirraway Interceptor. The CAS was directed to examine the delivery prospects for Kittyhawks and to review the Interceptor's performance. The Director General of Aircraft Production should assess whether the resources required for the fighter could be better used to produce other aircraft. However, the production of a single prototype was approved.

CAC's proposed Australian-built fighter had reached a critical point. The Interceptor's fate was resting on the advice War Cabinet was about to receive from the CAS and the Director General of Aircraft Production.

The future of the Wirraway Interceptor was only one of many issues facing the War Cabinet in February 1942, as it considered how to use Australia's sparse military and industrial resources to respond to a series of horrendous military reverses. Japanese forces landed in Malaya on 8 December 1941 and quickly routed its defenders, including the 8th Australian Division. The surviving troops had retreated to Singapore by 31 January 1942. Japan attacked Singapore on 8 February and more than 130,000 troops, including 15,000 Australians, surrendered on 15 February. In Australia's northern approaches the Japanese conquest of the Netherlands East Indies began in January 1942 and was well-advanced by February.

In a chilling demonstration of the flimsiness of the RAAF's fighter defence, eight Wirraway "fighters" that opposed a numerically superior Japanese air attack on Rabaul in New Guinea on 20 January 1942 were slaughtered. In less than ten minutes, three Wirraways were shot down, two crash-landed, and one crashed trying to take-off. Of sixteen crew, six were killed and five wounded or injured. Facing these disastrous developments, the War Cabinet surely would have welcomed straightforward, consistent advice about whether to proceed with the Fighter Interceptor project. However, it was not allowed that luxury.

The War Cabinet considered the additional information it had sought when it met again on 18 February 1942. The Aircraft Advisory Committee provided the report about the utilisation of aircraft production resources. The report was based entirely on information provided to the committee by CAC. It argued that Wirraway production was nearing completion and construction of the Interceptor would be an effective way to utilise CAC's 2,000 strong Wirraway workforce until it was redirected to production of the new CA-4 bomber. CA-4 production was scheduled to commence in September. It was claimed these men could not be redeployed to work on other aircraft types. They would have to be dismissed at a rate of 200 per week, dispersing a highly trained workforce. The report also argued that if the Wirraway Interceptor was not used as a fighter it could be used by operational training units.

CAS Burnett provided conflicting advice. Burnett firstly addressed the supply of fighters from overseas and concluded that Kittyhawks and Beaufighters would be supplied to Australia in early 1942. He then reviewed the performance of the Wirraway Interceptor and accurately concluded that:

> The estimated performance is insufficient to warrant the production of this aircraft except in the absence of modern fighters … and … the Wirraway Interceptor should only be produced as a makeshift pending the procurement of modern aircraft.

Burnett also pointed out that the Interceptor would be of little value for operational training units, which should provide trainees with experience on the aircraft they would use on operations.

Burnett concluded by recommending that three Wirraway Interceptors be produced to carry out service trials:

> … so that in the event of Kittyhawks or other modern fighters not coming to hand, then a production order for Wirraway Interceptors can be put in hand immediately.

Burnett's report to the War Cabinet used the dry, unemotional language expected when the RAAF's senior officer provided advice to his government. A more direct insight into Burnett's thinking is provided by his blunt comments to the Deputy CAS recorded on an RAAF file:

> I am no more convinced as to the necessity for building this aircraft than I was previously when the position of fighter aircraft appeared more difficult to supply. I cannot believe that two thousand men are going to be turned out of work due to inability to employ them on the CA-4 after work on the Wirraway has been completed.

The War Cabinet considered the conflicting advice from the CAS and the Aircraft Advisory Committee, which had effectively acted as a conduit for CAC and Wackett. It decided that 100 Wirraway Interceptors should be produced to utilise CAC's Wirraway workforce until it was allocated to production of the CA-4 Bomber and to provide fighters in case imported aircraft failed to arrive. The views of CAC had prevailed over the advice of the CAS. Australia would have a locally designed and produced fighter.

A newly assembled P-40E fighter at RAAF Amberley, Queensland, in January 1942. Within the next few months hundreds of P-39 and P-40s had arrived in Australia for use by both the USAAF and RAAF.

CHAPTER 4
TESTS AND FURTHER ORDERS

Shortly after approving production of 100 Wirraway Interceptors, the War Cabinet made an in-principle decision to expand the RAAF to 73 squadrons, including no less than 24 fighter squadrons. The harsh realities of war had shown that a strong fighter component is an indispensable part of a balanced air force. More fighters were urgently needed if this ambitious plan was to be achieved. For a second time, CAC appeared to have a ready-made solution.

Daniel McVey, Secretary of the Department of Aircraft Production, wrote to the Department of Air in July. He relayed advice from CAC that:

> … 80% of the fabrication work associated with the order for 100 "Boomerang" Interceptor Fighters has been completed and … unless an order for additional aircraft of this type is forthcoming shortly it would be necessary to divert the staff concerned to other work if it could be found for them or, alternatively, to dismiss them until such time as the production of the CAC Dive Bomber [the CA-4] is in full swing.

McVey asked whether further orders for the Boomerang were likely. However, further orders would depend on evidence the Boomerang could successfully compete with opposing fighters. The inescapable need for fighters to engage enemy aircraft on competitive terms had been demonstrated repeatedly during the war. To assess the Boomerang's performance, it was flown against American P-39 Airacobra and P-40 Kittyhawk fighters in a series of mock combats in July. The findings of these tests were assessed by the Air Board in early August 1942. The Board concluded that the Boomerang could dictate the terms of combat to the two American types, with this ability progressively increasing as altitude increased above 10,000 feet. The Boomerang could be used to bolster the RAAF's fighter strength. The Board would support an order for a further 100 Boomerangs, taking the RAAF's total acquisition of the type to 200 aircraft.

Remarkably, the Air Board reached these conclusions despite the tests showing the Boomerang was significantly slower than the Airacobra and the Kittyhawk, and both American aircraft could terminate combat at will by diving away at speeds the Boomerang could not match. The Air Board's willingness to overlook these adverse findings reveals a startling lack of objectivity in its assessment of the mock combats.

The Board did inject a note of realism by recommending that improving the performance of the Boomerang by increasing engine power should be a priority.

Despite the discouraging findings that the Boomerang could not match American P-39 and P-40 fighters in top speed or diving speed, the Air Board still concluded that:

> … under combat conditions, as far as these can be reproduced in practice, the Boomerang aircraft, due to its superior rate of climb and manoeuvre, can, in general, dictate terms of combat to both these American types.

The Board reached this conclusion after reviewing reports of the tests and speaking to participants. However, a US Army Air Force report of the same tests provides an interesting comparison with the Air Board's assessment. While not opposing production of the Boomerang, the USAAF report included the important observation that:

> It can readily be seen both the Kittyhawk and the Airacobra have the advantage over the Boomerang, due to their ability to terminate combat by diving away from the Boomerang.

Similarly, in the context of the Boomerang's possible employment against the Japanese Zero fighter:

> … it is believed that the Boomerang, once committed to action, cannot be withdrawn, due to lack of any characteristic which can be utilised to terminate combat … once committed there is absolutely no termination of combat.

These striking reservations about the Boomerang's air combat performance were notably absent from the Air Board's report.

The Air Board was responsible for leading and directing the RAAF. However, the Air Board failed to provide appropriate leadership during its evaluation of the July 1942 comparison tests. The Board did not make the same realistic judgement about the Boomerang's shortcomings as a fighter aircraft that CAS Burnett had made six months earlier.

There was a postscript to the Air Board's review of the comparison tests. A meeting between Boomerang test pilot Ken Frewin and senior RAAF and aviation industry figures on 10 September 1942 provided another opportunity to assess the Boomerang's performance. Participants included CAS George Jones (who chaired the Air Board), Director-General of Aircraft Production Essington Lewis, Secretary of the Department of Aircraft Production Daniel McVey, RAAF Group Captain Ellis Wackett (brother of Lawrence) and US Lend Lease representatives. The meeting had been precipitated by a letter from Frewin to the Minister for Aircraft Production Senator Donald Cameron. In the letter Frewin complained about the Boomerang's high-pressure flight-testing regime and limitations as a fighter, and what he saw (with good reason) as the failings of the CA-4 bomber project.

Frewin debated the merits of the Boomerang at the meeting. Frewin then withdrew and staff from the RAAF and Australia's aeronautical research agency who had taken part in the mock combats discussed the interpretation of these tests. The comments from the participants were revealing. They stated that:

CAC test pilot Ken Frewin in the cockpit of a Boomerang as Laurence Wackett looks on. Frewin was a former commercial pilot. (AWM)

… they [the results of the mock combats] do permit drawing general conclusions. These were to the effect that the CA-12 [the Boomerang] is slower than either the P-40 or the P-39. It has a greater rate of climb at all altitudes but this in itself is not the whole picture. By reason of their greater speed and particularly their ability to dive at greater speeds, the P-40 and P-39 can, by diving away and gaining great speed, reach certain altitudes faster than would be indicated by the normal rate of climb. They stated that the "Boomerang" is more manoeuvrable than either the P-40 or the P-39 but this characteristic would be an advantage only if the P-40 or P-39 were to elect to engage in combat. The great disadvantage of the Boomerang is that, because of its low speed, it could not get away in the event that the fight should turn against it.

These critical comments were consistent with those made by the USAAF. At the September 1942 meeting the truth about the Boomerang's shortcomings as a fighter were put directly to the most senior figures in the RAAF and the Australian aircraft industry. In response, CAS Jones informed the Advisory War Council about the meeting and blithely concluded that "there was no substance in Mr Frewin's statements".

Australia's wartime bureaucracy now moved to translate the Air Board's surprisingly favourable assessment into an order for additional Boomerangs. A further meeting of the Air Board on 28 August resolved to seek War Cabinet approval to increase Boomerang production to 200 and for projects to develop more powerful engines. The Aircraft Advisory Committee dutifully endorsed these recommendations on 1 September. Committee Secretary Daniel McVey wrote to Wackett on 2 September, stating CAC could increase Boomerang production to 200 and initiate projects to develop more powerful engines on the assumption that War Cabinet approval would be secured. In October the Ministers for Air and Aircraft Production recommended the additional Boomerang production to War Cabinet. The submission also proposed that three higher-powered prototypes be constructed, utilising a higher power (1,700 horsepower) Wright Cyclone engine, a turbo-charged version of the Pratt and Whitney Twin-Row Wasp and a Twin-Row Wasp with capacity increased from 1,830 to 2,000 cubic inches. The War Cabinet approved the Ministers' recommendations on the assumption that production could proceed without prejudicing other aircraft construction programs.

There was a revealing postscript to the War Cabinet's decision. It was referred to the Advisory War Council for endorsement, which was duly provided. However, CAS Jones reported that getting this endorsement was not straightforward. He attended the Council meeting and had to justify the additional production to some members of the Council, who questioned whether it was needed or whether fighters could instead be sourced from the United States or Britain. Fortunately for Jones, he had a powerful ally at the meeting. Jones recorded:

The Prime Minister did realise that lack of work and consequent dispersal of CAC staff would be regrettable, to say the least of it, and, as I understand it, he finally agreed to approve the order [for additional Boomerangs].

CAC's argument about the need to continue Boomerang production to avoid disrupting its workforce again overcame any opposition.

With this approval to increase production from 100 to 200 aircraft and to construct higher performance prototypes the future of Australia's local fighter looked encouraging. However, shortly after this moment of triumph, the seemingly high-flying Boomerang project encountered severe turbulence.

The War Cabinet called for a report about aircraft production policy when it approved increased Boomerang production. The direction to prepare this report stemmed from a wider concern about ensuring Australian industry could produce aircraft of equal or superior performance compared with their enemy equivalents. One approach would be to improve the performance of types already being produced in Australia, such as the Boomerang and the Beaufort. Alternatively, the best types of aircraft being produced by the Allies could be identified and approval sought to manufacture them in Australia.

The aircraft production policy report was prepared by CAS Jones and Director General of Aircraft Production Essington Lewis. Jones and Lewis recommended that Australia's aircraft production program be based on a prolonged war (1944 and beyond) and should include production of the best proven Allied fighters and bombers. They specifically considered the possible development of the Boomerang and shrewdly pointed out that:

> ... the success of the improved Boomerang would have to be demonstrated by test flying of the prototype [i.e. the version fitted with a 1,700 horsepower Wright Cyclone engine] ... these developments and trials ... would inevitably mean retardation of production and perhaps the taking of chances with our production set up.

They did recommend that development of the best-possible Australian fighter should proceed, but this was a small concession. It could not disguise a fundamental shift away from relying on Australian expertise to accepting the reality that designs should be sought overseas. When the War Cabinet approved the report on 7 December 1942, the Boomerang was effectively consigned to the margins of Australia's war effort.

The Boomerang again came to the notice of the War Cabinet in July 1943. This was triggered by a familiar refrain, from a familiar source. Wackett had written to the Department of Aircraft Production in June, stating that manufacture of Boomerang components was nearing completion. Boomerang assembly would be completed by October 1943. After this, employees allocated to Boomerang production would lack work until CAC commenced licence-production of the superb North American P-51 Mustang fighter in June 1944. This again raised the prospect of a valuable, highly trained group of workers being dispersed. This looming crisis could be alleviated by an order for a further 200 Boomerangs. Wackett further suggested these could be higher-powered versions.

Perhaps sensing a changed attitude to the Boomerang in the War Cabinet, Aircraft Production Minister Cameron presented a more modest recommendation that 120 additional Boomerangs be ordered. Cameron argued this would maintain CAC's manufacturing organisation until it started to produce the Mustang fighter. Mirroring Wackett's position, Cameron also flagged the possibility of the production run including higher-performing Boomerangs fitted with a turbocharged 2,000 cubic inch Twin Wasp engine.

But this time the War Cabinet had run out of patience with the Boomerang. War Cabinet noted advice from the CAS that no further Boomerangs were required by the RAAF, but additional Wirraways could be utilised. However, reinstating Wirraway production would be impractical. The War Cabinet therefore directed that:

(i) production of previously ordered Boomerangs be slowed down and the Mustang project be expedited as much as possible.

(ii) an order for a 50 additional Boomerangs be placed to maintain CAC's production organisation until Mustang production commenced.

The Boomerang was officially deemed unwanted by the RAAF and further production was grudgingly endorsed for the sole purpose of maintaining CAC's fighter production regime.

The War Cabinet considered the Boomerang on two further occasions. Neither was flattering for Australia's home-grown fighter. In September 1943 it belatedly agreed to fund the additional 50 Boomerangs it had approved in July. While approving this expenditure:

War Cabinet directed that it be recorded that it views with concern the further manufacture of these obsolescent aircraft.

Finally, in December 1944 the War Cabinet considered whether the last few Boomerangs should be completed or whether to use the components for these aircraft as spares. The War Cabinet agreed to the completion of all aircraft, relying on CAC's advice that this would assist in a smooth transition to the Mustang project. To leave its views in no doubt, War Cabinet directed that Boomerang production must cease by 31 January 1945. The Australian government's patience with the Boomerang was finally exhausted.

The Boomerang was ordered into large-scale production in February 1942 at a time of apparently existential crisis for Australia. However, the Boomerang faced further obstacles before it could enter service with the RAAF. Ordering the Boomerang was one thing, but CAC would find that turning orders into speedy mass production would be quite another.

Australian assembled Mustangs at the CAC factory in 1945. The aircraft on the left is A68-1 which first flew on 29 April 1945. The final batch of Boomerangs was completed to keep the factory busy pending the start of Mustang production. (AWM)

The maiden flight of the Boomerang took place on 29 May 1942 in front of a large crowd of CAC factory workers and their families. (AWM)

Boomerang prototype A46-1 which flew air combat trials versus a P-39 and a P-40 in July 1942. In early 1943 this aircraft entered service with No. 2 Operational Training Unit. (Michael Claringbould)

CHAPTER 5
THE PRODUCTION BATTLE: EARLY VICTORIES AND LATER LOSSES

The War Cabinet approved production of 100 Wirraway Interceptors on 18 February 1942. Even before this decision CAC and the RAAF continued the design process that started in December 1941, correctly anticipating government approval for large-scale production. Construction of a mock-up was well advanced by late January, using Wirraway sub-assemblies to the maximum extent possible.

A critical step was a design conference held at CAC on 11 February. At this meeting the RAAF's requirements for the new fighter were identified, so they could be included in a formal RAAF specification. Participants included Wackett, project Chief Engineer Bill Air and a whole gathering of RAAF specialist officers, who were responsible for aspects such as armament, signals, the electrical system, oxygen equipment, instrumentation and armour plating. The detailed technical requirements agreed at the conference were included in the document *RAAF Developmental Specification No.1/42 for Wackett Interceptor* issued on 6 March. The Department of Aircraft Production wrote to CAC on 16 March 1942, advising the company that production of 100 aircraft in accordance with this specification had been approved and:

> … this letter may be regarded as the official order to proceed with the manufacture of the abovementioned aircraft.

By April the RAAF had formally issued *Contract Specification No. 4/42 for Aircraft Type CA12* to CAC.

Work on the factory floor continued in parallel with the formal procurement process. By early March completion of all the jigs used to manufacture the aircraft was imminent and production of fuselages was underway. Wind tunnel testing of a model commenced on 10 March. The extreme pressure to produce a fighter to defend Australia was felt by everyone. Fred David later recalled that:

> Everyone had the wind-up. It was a very precarious situation so we just got cracking … what occurred showed the resourcefulness of Australians under pressure … I think I worked about 70 hours a week at that time.

Despite all the efforts to expedite production of the Interceptor Fighter, the difficulties associated with designing and producing a combat aircraft in Australia in 1942 sometimes led to unusual improvisations. Browning 0.50-inch calibre (12.7mm) machine guns had been identified as a potential weapon for the Interceptor. The RAAF endeavoured to locate examples of these weapons to examine their possible integration with the Interceptor. Cooperation between the RAAF and the USAAF led to a RAAF officer accompanying a US airman to the Montague train station in suburban Melbourne in February to borrow two US Browning 0.50-inch guns, which were being stored at this distinctly non-military location.

The weapons were being kept at the station pending their installation on US P-40s already in Australia. The guns were delivered to CAC on the understanding they would be returned to US forces when required, which was estimated to be about ten days. Despite this valiant improvisation, the Boomerang did not incorporate 0.50-inch Brownings.

The need for a more powerful armament to supplement the fighter's rifle-calibre 0.303-inch machine guns was well recognised. CAC attempted to manufacture 20mm cannon based on an example brought back from North Africa by an Australian serviceman. Large-scale production was never achieved, despite much hard work that resulted in the manufacture of a few examples. Fortunately, British-made Hispano 20mm cannon soon became available in substantial numbers. The Boomerang's armament included two 20mm cannon and four locally produced 0.303-inch calibre machine guns.

On 29 May 1942 the hard work of hundreds of CAC staff from all levels of the corporation reached a thrilling milestone when the first Boomerang lifted off at CAC's factory airfield in the hands of test pilot Ken Frewin. For a nation engaged in total war, where absolute secrecy about military technology would be expected, the event was surprisingly public. To Frewin's consternation a crowd of thousands was present and the aircraft was on display. The wives and children of CAC directors joined in the festivities. Fortunately, the crowd was treated to a thoroughly successful test flight. Frewin found that the Boomerang took off readily, was easy to fly and landed in a short distance.

Some of Frewin's subsequent test flights were more eventful. While demonstrating the aircraft for the CAS, the cockpit canopy tore loose and struck Frewin on the head during a 480 kilometre per hour low-level pass, fortunately without doing much harm to the pilot. Frewin was particularly upset because his previous recommendation to strengthen the cockpit structure had been ignored. The trailing-edge of one blade of a wooden test propeller broke off during a subsequent flight, resulting in three engine bearings breaking, the pitot tube (used to measure air speed) being smashed and the tailplane spar breaking. Frewin used all his skills to make a forced landing at Werribee airfield near Melbourne.

While these early test flights were underway a development of a different type occurred in early June 1942. CAC named the new fighter the Boomerang in accordance with a recommendation from the Air Board.

The battle to design and fly the Boomerang had been won by mid-1942, reflecting the extraordinary efforts of CAC's workforce. In June 1942 the Advisory War Council received the exciting news that:

> Tests of the Commonwealth Aircraft Corporation fighter had proved satisfactory. This aircraft has a speed of 315 mph [507 kmph] with great manoeuvrability and ability to climb … mass production would be commenced immediately.

This had been a hard-fought and justly celebrated achievement, but an even more arduous struggle was about to begin.

Australia's complete lack of modern fighters at the start of the Pacific War in December 1941 was the impetus for the Boomerang's development. Although designing, constructing and flying the first example were important steps, the crucial next task was mass producing the aircraft to enhance the RAAF's fighter strength.

Early statements about production from Wackett and CAC had been extremely optimistic. As noted earlier, Wackett wrote to aircraft production chief Essington Lewis on 31 December 1941 promising that (with sufficient engines and propellers being available) the first aircraft could be delivered within three months and a total of 200 within a further six months, i.e. 200 aircraft would be delivered by September 1942. Similar numbers continued to be promoted in early 1942. In January, the short-lived Aircraft Production Commission stated that:

> We can be in production by June and complete 200 in six months.

In February Wackett reinforced his promises for speedy production, promising that:

> Delivery of the first of these Interceptors can actually be made in three months' time, and fully 100 can be delivered in a short period of three months.

These expectations of rapid production were formalised when the Department of Aircraft Production confirmed the government's order for 100 aircraft on 16 March 1942. The confirmation stated that:

> Delivery of these aircraft is to be made … as soon as possible, it being understood that deliveries will commence not later than June 1, 1942, and will be completed by September 30, 1942.

Wackett's optimism about production of the Boomerang continued when he discussed the aircraft with Lieutenant General George Brett, USAAF, on 8 June. Brett held the very important post of Commander, Allied Air Forces, South-West Pacific. He recorded that:

> I was informed by Mr Wackett that he could have 25 to 30 of these ready by the 1st of August, 100 by the 1st of October and an average production of 45 a month thereafter.

Brett's report also raised the:

> Possibility of re-arming the [RAAF's] 75th Squadron [with the Boomerang] complete about the 1st August and getting it back into operation as soon as possible; to then re-arm the 76th and 77th with the Wackett Interceptor.

However, initial expectations for speedy production and introduction to RAAF service ran into hard realities as 1942 progressed.

These difficulties are starkly illustrated by CAC delivery schedules, which specify the corporation's expected dates for delivery of completed Boomerangs to the RAAF. CAC prepared several of these schedules during 1942 and 1943. Each schedule shows significant delays in Boomerang deliveries compared with its predecessor. The planned delivery dates in the schedules are best shown via a chart, which also shows actual deliveries to the RAAF:

In marked contrast to CAC's overly optimistic schedules, the 250th and last aircraft was actually

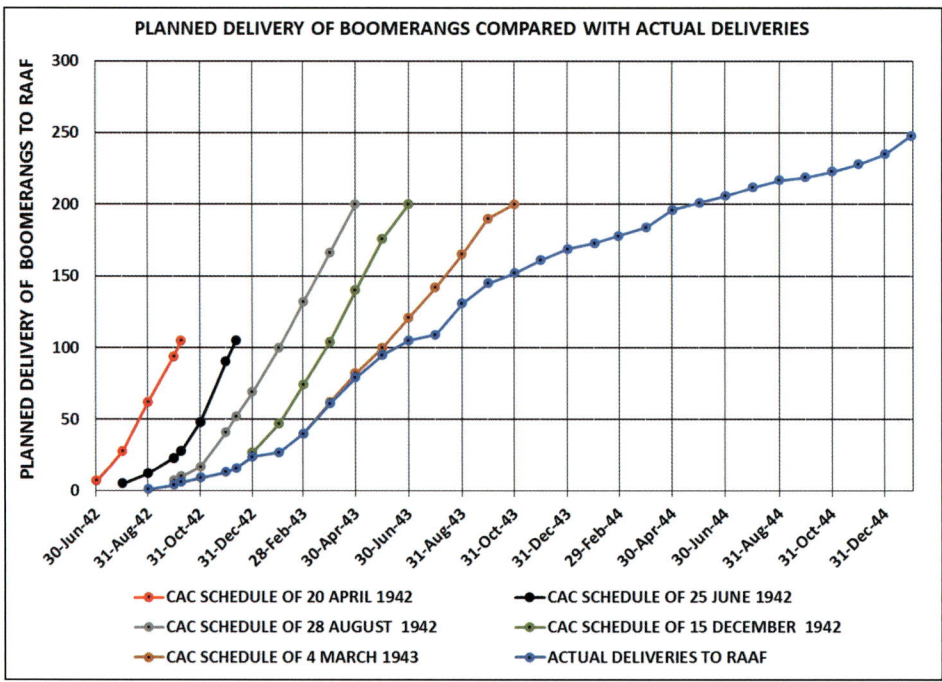

Planned and Actual Deliveries of Boomerangs to the RAAF 1942-44

delivered to the RAAF on 1 February 1945. Indeed, Boomerang production never went close to meeting the initial company forecasts. With hindsight this is unsurprising, given the problems faced by CAC as it is scaled up to mass production. The Boomerang was approved and ordered off the drawing board, and CAC had not previously attempted mass production of a fighter aircraft. Inevitably, adjustments were required after the Boomerang took to the air. Wackett explained to the Department of Aircraft Production on 26 June 1942 that:

> … we have had three weeks of test flying, and have encountered various difficulties. We are now able to review the effect of the necessary changes on our anticipated production. In addition, we have been able to allow for many contingencies which were not fully apparent.

When Wackett sent another schedule to the Department on 28 August 1942, he stated that:

> Various troubles which have arisen in getting production under way have been taken into account … With five aircraft completed, we are for now able to make the best estimate which is possible and to allow for lags which have occurred.

Another problem was identified by CAC in December 1942. The Boomerang was powered by Twin Wasp engines imported from the United States, but these imports were suffering delays. These delays resulted in some completed airframes being stored pending supply of engines. Sluggish delivery of Twin Wasps was still slowing Boomerang production as late as March 1943. These setbacks pushed production into 1943, when the inevitable competing demands for resources associated with a total war compounded the earlier delays.

Boomerangs under production at CAC, circa 1943. (AWM)

By 1943 Australian industry was undergoing unprecedented mobilisation to meet the needs of a war economy. This mobilisation brought about unrelenting demands for labour, which was always in short supply. The labour shortage extended to CAC and Boomerang production. CAC advised the Department of Aircraft Production on 4 March 1943 that:

> … the manpower position in relation to production has shown no appreciable improvement, and … the position has now been arrived at when it can be stated with some certainty that the 200th Boomerang aeroplane will not be finished until October.

The labour shortage was exacerbated by dwindling government support for the Boomerang. In July 1943 the War Cabinet reluctantly approved production of an additional 50 aircraft and in September of that year it condemned the Boomerang as obsolescent. Prime Minister (and Minister for Defence) John Curtin wrote in blunt terms to Minister for Aircraft Production Donald Cameron on 9 September 1943, noting that the Boomerang was still listed as an "Absolute Priority" project. Curtin demanded that Cameron observe the War Cabinet's decision to slow Boomerang production and divert all possible resources to the Mustang project. Facing labour shortages and the relegation of Boomerang production to secondary status, grudgingly tolerated to maintain CAC's fighter production regime, the final 100 or so aircraft trickled off the production line from October 1943 to January 1945.

CAC could be proud of moving from initial design to first flight in a mere five months but implementing mass production proved far more challenging. Deliveries of aircraft to the RAAF never came close to meeting the early, absurdly optimistic forecasts. The inevitable teething problems associated with such a rapidly developed type caused delays. This meant that production timelines blew out until labour shortages were encountered, causing further hurdles. Production continued merely to utilise CAC resources until they could be redeployed to other programs. The story of the Boomerang's design and production started with a bang but ended with a whimper.

Boomerang A46-122, coded as No. 83 Squadron aircraft MH-R. (Author)

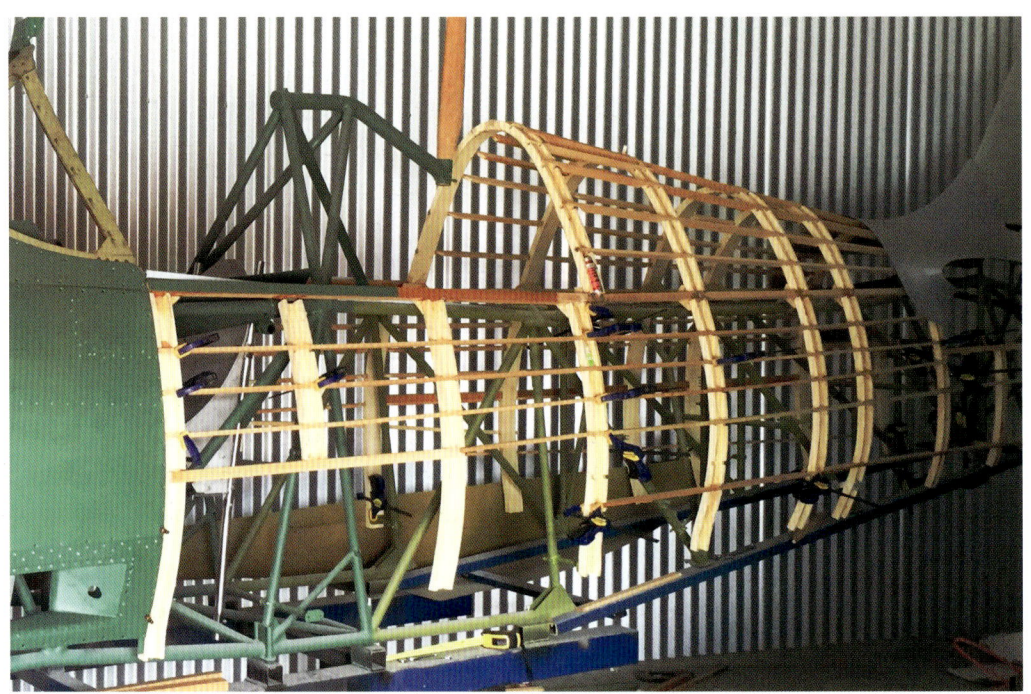

The rear fuselage of A46-89 during restoration. Wooden formers (vertical components), wooden stringers (horizontal components) and the steel tubular framework are visible. (Ian Baker)

CHAPTER 6
THE CAC BOOMERANG DESCRIBED

The CAC Boomerang was a single-seat, single-engine fighter. This type of aircraft was the key to dominating contested airspace in WWII. The Boomerang incorporated a fuselage with a steel tubular frame, aluminium stressed-skin wings and a cantilever tail unit (empennage). It was derived from the Wirraway but included many original elements.

The fuselage was a hybrid design and requires careful description. It was shorter than the Wirraway's and was based on a chrome-molybenum steel tubular framework. This framework was a two-piece unit that included separate forward and rear fuselage frames. From the cockpit back to the rudder, the external top and sides of the fuselage consisted of a plywood shell fitted over wooden formers and stringers. This wooden construction replaced the fabric-covered aluminium frame used in the corresponding parts of the Wirraway fuselage. The plywood shell was attached to an aluminium semi-monocoque stressed-skin lower fuselage section by screws.

The metal-tube based construction of the fuselage was shared with the earlier North American NA-16 trainer and the Wirraway. However, this technology was not as advanced as that used in front-line fighters of the early 1940s, such as Britain's Supermarine Spitfire, Germany's Messerschmidt Bf 109 and Japan's Mitsubishi Zero. These fighters incorporated all-metal, stressed skin monocoque fuselages. This more modern construction technique reduced weight while increasing strength. However, the urgent need to develop a fighter for Australia's defence in early 1942 meant that the Boomerang's core design and construction techniques mirrored those of the Wirraway. Accordingly, the Boomerang did not incorporate the most modern construction methods available at the time.

Forward of the cockpit, the installation of the more powerful Twin Wasp engine required various changes compared to the Wirraway design. These included a strengthened forward fuselage to cope with the increased engine power, stronger engine mounts and moveable gills at the rear of the engine compartment to provide an adequate flow of cooling air under all operating conditions.

The single-seat cockpit was located over the centre of the wing and was enclosed with a sliding canopy. The cockpit featured armour protection and a 38mm bullet-proof windscreen. The pilot was provided with a set of instruments typical of fighter aircraft of the period.

The low-mounted wings were of the cantilever type and included five sections. These were the centre section, an outer panel in each wing and the wingtips. The centre section incorporated two spars and was of constant chord. Although it had the same dimensions as the Wirraway equivalent, the Boomerang's centre section was strengthened to cater for the additional stresses associated with the Boomerang's greater weight and speed, recoil from the cannons and carriage of a drop tank under the wing centre section.

A46-122 being serviced at the Temora Aviation Museum. Visible are the Twin Wasp Engine, open cooling gills immediately behind the engine and straight-through engine exhaust. Part of the fuselage tubular framework at the tail can also be seen as well as the location of the join between the wing centre section and the outer wing section. (Author)

The Twin Wasp engine in A46-122, showing the two rows of cylinders. (Author)

The all metal, stressed-skin outer wing sections included a single spar. They were significantly shorter than the Wirraway's but did incorporate many of the two-seater's design features, such as swept back leading edges, straight trailing-edges and tapering thickness. The fabric-covered ailerons were fitted with wooden trim tabs. The split trailing-edge flaps were located between the ailerons and the fuselage. The initial CA-12 aircraft were fitted with metal wingtips, but later aircraft had wooden units. The tail was generally of metal skin construction, but all moveable surfaces were fabric covered and fitted with aluminium trim tabs.

The undercarriage included mainwheels that hydraulically retracted inwards to wheel wells located in the centre wing section, forward of the front spar. The tail wheel was non-retractable and semi-steerable and was full-castoring.

Fuel (100 octane) was carried in one fuselage-mounted self-sealing 70-gallon tank (318 litres) and two 45-gallon (204 litre) tanks located in the wing centre section. The wing-mounted tanks had the same capacity as those installed in the Wirraway, but the fuselage tank was new to the Boomerang. A 70-gallon (318 litre) plywood drop tank (referred to as a belly tank by the RAAF at the time) could also be carried under the wing centre section. Total internal capacity was 160 gallons (726 litres) or 230 gallons (1,044 litres) with the drop tank.

The Boomerang was derived from the Wirraway, but much work was required to convert a training/general purpose aircraft into a fighter, which was more compact, fitted with a much more powerful engine and designed to cope with the greater stresses associated with aerial combat. A very significant proportion of the Boomerang's parts were newly designed, confirming that the aircraft was largely original. The Boomerang was far more than just a modified Wirraway.

The Boomerang was powered by a Pratt and Whitney R-1830-90B Twin Wasp S3C4-G 14-cylinder double-row radial with a displacement of 1,830 cubic inches (30 litres). The rear case of the engine incorporated a single-stage, two-speed centrifugal type supercharger. Important engine and propeller specifications were:

Take-off power = 1,200 horsepower at 2,700 RPM

Power with Military Rating (2,700 RPM) = 1,200 horsepower at 4,900 feet and 1,050 horsepower at 13,500 feet.

Propeller: three-blade De Havilland Type 3E50, hydraulic operation, diameter = 11 feet (3.35 metres)

Until the late-war appearance of jet and rocket-powered aircraft, all WWII fighters were powered by piston engines that drove propellers. In this context a few observations will be made in respect to the Boomerang's engine.

A crucial task for aircraft designers was to combine the most powerful available engines with suitable airframes. This maximised power to weight ratios, which in turn maximised performance in critical areas such as top speed, acceleration and rate of climb. The most powerful engine available to CAC for its Interceptor Fighter was the Pratt and Whitney R-1830 Twin Wasp radial engine.

The radial engine was one of two basic WWII aircraft engine configurations, the other being the inline. The cylinders of a radial engine are arranged radially around the crankshaft in a single plane, similar to the arrangement of bicycle wheel spokes around the axle. Early aircraft radials included a single row of cylinders, but the R-1830 Twin Wasp was an example of the "twin-row" radial, which included two rows of cylinders. This configuration increased engine capacity and power without increasing diameter. WWII radials were almost always air cooled. In contrast to radials, the cylinders of inline engines were arranged in straight lines. These were invariably liquid cooled and required radiators.

The design philosophy of installing the Twin Wasp in an airframe derived from the Wirraway training/general purpose aircraft fundamentally influenced the technical characteristics and performance of the Boomerang.

CAC's engine factory at Lidcombe in Sydney manufactured the Twin Wasp under licence from Pratt and Whitney, the first run of its first engine taking place in September 1941. The factory was built to produce Twin Wasps for the Australian-made Beaufort bomber. With the demand for Twin Wasps for the Beaufort exceeding the rate at which Lidcombe could produce them, no locally manufactured engines were available for the Boomerang. Additionally, Lidcombe produced the single-speed supercharger version of the Twin Wasp (S1C3-G), which suffered a performance drop at high altitude, compared with the two-speed supercharger Twin Wasp (S3C4-G) installed in the Boomerang.

Fortunately, US production solved this problem, as it did for many types of WWII armament production. Large numbers of Twin Wasps were imported from the US and Boomerangs were fitted with imported S3C4-G variants of the engine. However, even the cornucopia of US production had its limits early in the Pacific War and insufficient supply of imported S3C4-G engines was one of the factors that delayed Boomerang production.

The installation of the Twin Wasp radial engine in a short airframe meant that the Boomerang had a squat, snub-nosed appearance. The aircraft was quite small compared with its peers. Its dimensions were:

Wingspan	36 feet 0 inches (10.97 metres)
Overall length	25 feet 6 inches (7.77 metres)
Overall height	9 feet 7 inches (2.92 metres) tail down, one blade vertically downwards; 12 feet 4 inches (3.76 metres) tail down, one blade vertically upwards
Gross wing area	225 square feet (20.90 square metres)
Dihedral	Centre section = zero
	Outer wing = 5 degrees and 30 seconds
Weights	Empty: 5,373 pounds (2,437 kilograms)
	Take-off with military load, without belly tank: 7,699 pounds (3,492 kilograms)
	Take-off with military load and belly tank with 70 gallons of fuel: 8,249 pounds (3,742 kilograms)

Armament	Two 20mm Hispano cannon (one in each wing), each with 60 rounds.
	Four 0.303-inch (7.7mm) Browning machine guns (two in each wing), with 1,000 rounds for each gun.

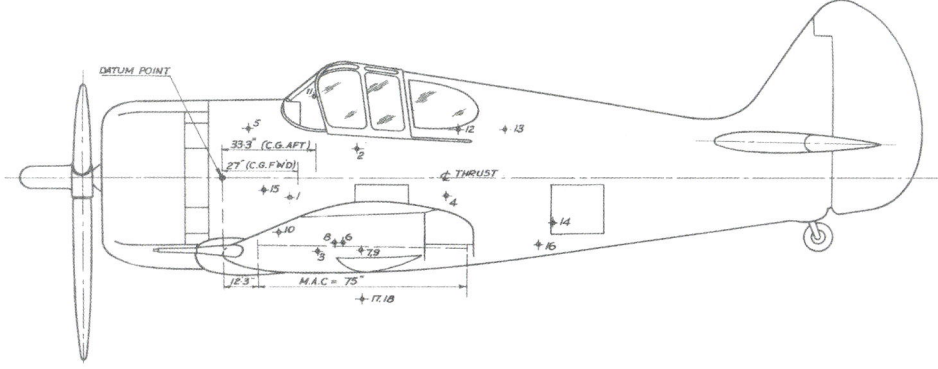

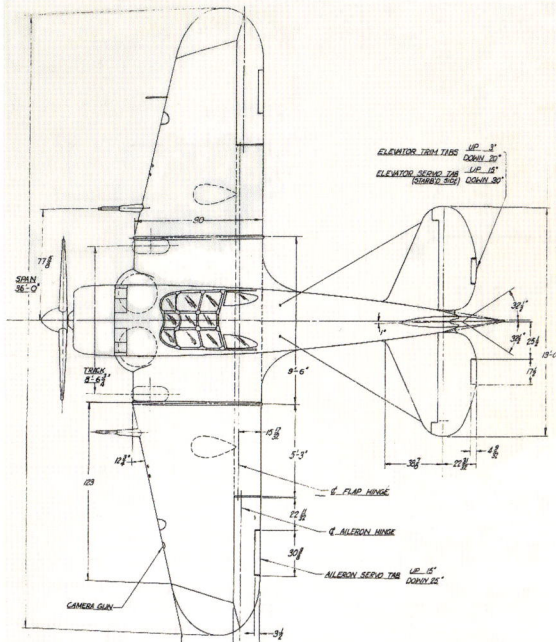

Above and left: Line drawings of the Boomerang from port side and overhead, reproduced from the 1943 CAC publication Boomerang Interceptor Overhaul and Repair Manual.

Below: The cannon and machine gun bay in the port wing of A46-122. The location of the join between the wing centre section and the outer wing is also visible. (Author)

Performance

The comparative trials between a Boomerang, Kittyhawk and Airacobra mentioned earlier yielded the following performance data for the Boomerang:

Altitude, Feet	Speed	Rate of Climb, Feet Per Minute
Sea level	260 mph/418 kmph	2,500
5,000	280 mph/451 kmph	2,500
10,000	295 mph/475 kmph	2,300
15,000	295 mph/475 kmph	2,080
20,000	300 mph/482 kmph	1,550
25,000	285 mph/486 kmph	1,050
30,000	260 mph/418 kmph	500

Service ceiling:	34,000 feet	
Range:	930 miles (1,490 kilometres)	

These results show that the performance of the Boomerang was inadequate in terms of its designated role as an "interceptor fighter". While the Twin Wasp radial was the most powerful engine available in Australia at the time, it was underpowered by 1942 standards. The Boomerang did not have sufficient speed to either engage contemporary fighters in air-to-air combat on equal terms or, just as importantly, overtake and engage enemy bombers. This deficiency is confirmed by the following performance data for Japanese aircraft:

Aircraft	Role	Max Speed
Mitsubishi A6M3 Zero (Sakae 21 engine, in service 1942)	Fighter	336 mph (541 kmph)
Nakajima Ki 43-II Oscar (in service late 1942)	Fighter	320 mph (515 kmph)
Kawasaki Ki 61-I Tony (in service 1943)	Fighter	348 mph (560 kmph)
Mitsubishi G4M1 Betty (in service at start of Pacific War)	Bomber	265 mph (428 kmph)
Mitsubishi Ki 21-II Sally (in service at start of Pacific War)	Bomber	297 mph (478 kmph)

Boomerang Variants

Although the Boomerang did not undergo major changes during production several variants were manufactured. These can be identified by the internal contract numbers that CAC allocated to the successive batches it delivered to the RAAF. These were:

CA-12: Initial production batch of 105 aircraft

CA-13: Next batch of 95, with a number of detailed design changes, including:

Wooden instead of metal wingtips

Mechanical instead of hydraulic cocking of the 20mm cannons

Provision of an external rear vision mirror and a signal pistol in the cockpit

Starting with A46-126 (i.e. the 126th Boomerang produced and hence the 21st of the CA-13 variant) fitting flame dampers to replace the original straight-through engine exhaust.

The Department of Air determined that these aircraft would be known as the Boomerang Mk II and aircraft 1 to 105 would be identified as the Boomerang Mk I.

CA-14: This contract number relates to a single aircraft, which was modified by CAC during experiments with higher-powered turbo-supercharged engines. The aim was to significantly improve the aircraft's performance and create a truly capable interceptor fighter. During the testing regime the aircraft underwent a series of modifications. The later, more heavily modified version was referred to as the CA-14A. However, CAC shifted its attention to production of the superlative P-51 Mustang fighter and efforts to produce a high-performance Boomerang variant became redundant. This aircraft remained a one-off.

CA-19: Final 49 production aircraft. The eleventh CA-19 delivered, and subsequent aircraft could mount a Williamson F24 camera in the lower rear fuselage for photo reconnaissance duties. The aircraft in this batch retained the Mk II designation, as evidenced by RAAF aircraft record cards.

The camera is removed from CA-19 A46-222, a No. 5 Squadron Boomerang named Glamour Girl. The fighter is seen on Bougainville shortly before the end of the war in July 1945. (AWM)

Variants were usually identified by contract number. References to Boomerang mark numbers are not common. Boomerang variants can be summarised as follows:

Aircraft No. / Date of Delivery to RAAF	RAAF Designation	CAC Contract
A46-1 to A46-105 / July 1942 to June 1943	Boomerang Mk I	CA-12
A46-106 to A46-200 / August 1943 to May 1944	Boomerang Mk II	CA-13
A46-201 to A46-249 / May 1944 to February 1945	Boomerang Mk II	CA-19
A46-1001 / April 1943		CA-14

Flying the Boomerang

Flying the Boomerang was a much more direct, hands-on experience than flying a modern aircraft. This perhaps mirrors many experiences of the 1940s compared with their twenty-first century equivalents.

The hands-on nature of flying the aircraft is shown by the following Boomerang cockpit drill. This was prepared by Wing Commander Peter Jeffrey, commanding officer of the RAAF's No. 2 Operational Training Unit:

Cockpit Drill

1. ***Before Starting***

    Trim   (i)   *set elevator trim to neutral*
              (ii)   *set rudder trim to neutral*

    Mixture   (i)   *set mixture control to idle out (fully forward)*

    Pitch   (i)   *set pitch control in fully fine (fully forward)*

    Fuel   (i)   *Turn fuel cock to selected tank*
              (ii)   *Check all fuel gauges*
              (iii) *Check fuel warning light*
              (iv) *Turn carburettor air temperature switch to 1*

    Blower   (i)   *Blower in low ratio*

    Guns   (i)   *Check that all gun switches are in the "off" pos.*
              (ii)   *Check that cannon charging handles are fully out*
              (iii) *Check that cannon charging handles are at 12 o'clock*

    Undercarriage (i)   *Select "wheels down" position and pump hand hydraulic pump until SOLID*
              (ii)   *Return selector lever to NEUTRAL position.*

    Flaps   (i)   *Check flaps for correct operation*
              (ii)   *Return flaps to "UP" position and selector lever to NEUTRAL position*

    Gills   (i)   *Place cooling gills in "FULLY OPEN" position, unless quick warm-up required, then use half-open*

2. ***Starting***

              (i)   *Pull engine through if "COLD"*
              (ii)   *Open throttle approximately one quarter open*
              (iii) *Raise fuel pressure to 15 lbs and turn the electric fuel pump off*
              (iv) *Prime engine according to engine temperature half to four stokes with primer pump. Lock pump after priming.*
              (v)   *Energise inertia starter until maximum revolutions are obtained.*
              (vi) *Turn ignition switch to "BOTH"*
              (vii) *When all clear given, engage starter*

     (viii)   *Return mixture control to Auto Rich as soon as engine fires*

     (ix)   *Set throttle to minimum of 6-800 r.p.m. for 30 secs*

     (x)   *Switch generator main line switch to "ON"*

     (xi)   *Warm up at 1000 r.p.m. until oil temperature reaches 40 degrees*

3.   **Taxying**

     (i)   *Engine revolutions must not be raised above 1000 r.p.m. until minimum oil temperature of 40 degrees C is obtained. Oil pressure during warming up – 105 lbs sq. inch maximum.*

     (ii)   *Move to take-off position as soon as oil temperature reaches 40 degrees C*

     (iii)   *Do not wear out brakes by excessive use and fast taxi-ing*

4.   **Before take-off**

     (i)   *Face aircraft into the circuit 45 degrees out of the wind*

     (ii)   *Check oil temperature 40 degrees C min – 60 degrees C desirable*

     (iii)   *Run up engine to 30" boost and 2000 r.p.m. and check oil and fuel pressures*

     (iv)   *Check switches at 1500 r.p.m. and note any loss of r.p.m.*

     (v)   *Check for free operation of Fuel Selector Valve*

     (vi)   *Check position of trim tabs*

     (vii)   *Check Flying Controls for free movement*

5.   **Take-off**

     (i)   *Open Gill and Oil Shutters according to engine temperatures*

     (ii)   *Pitch control in fully fine (Fully Forward)*

     (iii)   *Mixture control in "AUTO RICH" position*

     (iv)   *Blower in "LOW" ratio*

     (v)   *Open throttle slowly*

6.   **After Take-off**

     (i)   *Retract undercarriage. When indicator levers show undercarriage up check by using hand pump. Hand pump level should be almost solid. Return undercarriage selector to "NEUTRAL".*

     (ii)   *Whilst undercarriage is retracting reduce manifold pressure and r.p.m.*

     (iii)   *At 1000 feet change from selected tank to belly tank if fitted*

     (iv)   *Adjust cooling gills to give desired temperature (204 degrees C – 232 degrees C)*

7.   ***General Flying Limited Conditions***

     (i)   *Military (5 mins) "LOW" Blower Max r.p.m. 2700 max Manifold Pressure 45" Hg.*

     (ii)   *Military (5 mins) "HIGH" Blower Max r.p.m. 2700 max Manifold Pressure 41" Hg.*

     (iii)   *Rated power "LOW" Blower 2550 r.p.m. Manifold Pressure 41" Hg.*

     (iv)   *Rated power "HIGH" Blower 2550 r.p.m. Manifold Pressure 40" Hg.*

     (v)   *Maximum Cruising "LOW" Blower 2550 r.p.m. Manifold Pressure 28" Hg.*

     (vi)   *Maximum Cruising "HIGH" Blower 2550 r.p.m. Manifold Pressure 29.5" Hg.*

     (vii)   *Oil Pressure Min 65 lbs Max 105 lbs Desirable 100 lbs*

     (viii)   *Maximum Oil Temperature 85 degrees C level 100 degrees C climb*

     (ix)   *Maximum cylinder head temperature 280 degrees C*

     (x)   *Cylinder head temperature - Continuous level flight 232 degrees C*

     (xi)   *Cylinder head temperature – Cruising 204 degrees C*

     (xii)   *Fuel pressure allowable 14-16 desirable 15 lbs*

8.   **Landing**

     (i)   *Test warning horn*

     (ii)   *Set Fuel Selector Valve on tank with most fuel*

     (iii)   *Set mixture control to "AUTO RICH" position*

     (iv)   *Set Blower in "LOW" ratio*

     (v)   *Set pitch control in fully fine (Fully Forward)*

     (vi)   *Landing gear in "FULL DOWN" and "LOCKED"*

     (vii)   *Landing flaps in "DOWN" position "LOCKED"*

The Boomerang's Unsung Virtues

While the Boomerang may have lacked the outright performance to succeed as an "Interceptor Fighter" (a view reinforced by a contemporary assessment included in the next chapter), it did possess a number of useful attributes. Test pilot Ken Frewin stated that:

> It is easy to fly. It has a very quick "take off", and lands easily in a small space. It should be ideal for … small, advanced aerodromes.

The aircraft was agile and manoeuvrable, making it well suited to low-level flight. While contemporary accounts written for home front consumption should be treated with caution, a war correspondent's 1943 description of Boomerangs carrying out tactical reconnaissance missions in New Guinea vividly stated that:

> The Boomerangs daily give exhibitions of low flying over the valleys that leave ground troops with the hair rising at the nape of the neck. You never see a Boomerang flying much higher than 50 feet above the floor of the valley or the peak of a mountain. The great proportion of their reccos are made at 40 feet.

The Boomerang was also ruggedly constructed, robust and reliable. This combination of qualities was very useful when the Boomerang was used to carry out army cooperation missions in support of Australian ground forces.

CHAPTER 7

THE BOOMERANG GOES TO WAR: RAAF SERVICE

Having considered the Boomerang's origin, production and technical specifications we can now examine its operational use by the RAAF in WWII.

Operational Training Units

The start of war in September 1939 created an enormous demand for trained aircrew. Australia initially focused on supporting Britain's war effort via its participation in the Empire Air Training Scheme. The training demand surged again when hostilities with Japan erupted in December 1941.

Training on the industrial scale demanded by a world war required an enormous system to take in raw recruits and turn out thousands of qualified, combat-ready aircrews. Under the system that evolved after the outbreak of war with Japan, a trainee pilot firstly received brief, ground-based training at an Initial Training School. The trainee then progressed to an Elementary Flying Training School (EFTS) where he underwent flight training with an instructor, typically in Tiger Moth or Gypsy Moth biplanes. The unsurpassed excitement of his first solo flight was followed by further flying exercises and evaluation. Successful EFTS graduates would proceed to a Service Flying Training School, where pilots deemed suitable for single-engine aircraft would fly the Wirraway trainer.

Finally, after a training regime approaching twelve months, pilots were posted to Operational Training Units (OTUs) that employed types used by front-line units. These included single-engine Kittyhawks and Spitfires, and multi-engine Beauforts, Ansons, Hudsons and Catalinas. Boomerangs served with No. 2 OTU and No. 8 OTU.

No. 2 OTU was formally established at Port Pirie in South Australia on 6 April 1942 before moving to Mildura in Victoria a month later. In September the unit's aircraft establishment was set at 36 Boomerangs, 30 Wirraways, 12 P-40 Kittyhawks and a single Avro Anson (but

A formation of Boomerangs, likely during service with No. 2 OTU in 1943-44. (AWM)

this establishment strength was not reached for some time, as Boomerang production was frustratingly slow and Kittyhawks were desperately needed by front line squadrons).

In late 1942 deliveries of Boomerangs to the RAAF finally commenced and examples were issued to the unit. The Boomerang made its initial contribution to Australia's war effort in the seemingly limitless, bright-blue skies typical of Mildura's semi-arid climate, honing the skills of pilots before they transferred to operational squadrons.

While serving at Mildura the Boomerang suffered the indignity of being severely criticised in a *Report on Boomerang Aircraft*. The report was prepared by the unit's commanding officer, Wing Commander Peter Jeffrey, who provided an unflattering critique of the Boomerang's suitability as a fighter. His blunt report included comments such as:

DIVING AND ZOOMING

In a dive the aircraft is hard to hold and skids to the left. The acceleration in a dive is slow compared with other types, and the maximum allowable airspeed is far too low. In pulling out from a high-speed dive, considerable squashing takes place, and the speed falls off rapidly. The Zoom from a dive is consequently poor. The aircraft buffets and vibrates excessively at speed, and the sights cannot be held steadily on a target.

COMBAT MANOEUVRES

The handling characteristics of the aircraft at high "G" are poor. In a steep turn at high "G" the stick develops a shudder similar to that felt in most aircraft at the stalling speed and is probably due to "tail buffeting". The buffeting is also apparent when pulling out of a dive and is accentuated in both manoeuvres when gills and/or oil radiator flaps are open.

When this tail buffeting occurs, the stick loses all feel, and it would be impossible to aim the aircraft accurately.

Jeffrey's conclusions included the following bleak assessment of the Boomerang's potential as a front-line fighter:

It is considered by this unit, that the Boomerang, as an operational fighter aircraft, has no feature to recommend it.

(i)   Its general handling characteristics are poor in comparison with other modern types.

(ii) Its all-round performance, in comparison with enemy fighter aircraft, is extremely poor.

(iii) Its all-round performance, in comparison with other Allied fighter aircraft, is also poor.

Jeffrey was well qualified to make these observations, having served as a fighter pilot with the RAAF's No. 3 Squadron in the North African campaign, where he achieved ace status with five victories. Jeffrey rose to command No. 3 Squadron and subsequently was wing leader of the RAF's No. 234 Wing. Known as a forthright, innovative leader, Jeffrey sent his report directly to the Air Board, which henceforth had no excuse for not understanding the Boomerang's limitations.

A pair of Boomerangs from a training unit, likely No. 2 OTU in Mildura. The unit commander, Peter Jeffrey, was highly critical of their performance. (AWM)

The Boomerangs serving with No. 2 OTU suffered from the hard landings, taxying accidents and ground loops expected when inexperienced pilots were trained hard to ready them for combat. Fortunately, there was only one fatal incident, but this resulted in two deaths. On 1 June 1944 Flying Officers Roger Byrne (A46-14) and Alfred Knapman (A46-44) suffered a mid-air collision while carrying out a gun camera exercise some 32 kilometres west of the unit's airfield. Both were killed.

The demand for fighter pilots inexorably increased as the war progressed. A fresh OTU, No. 8, was established at Narromine in New South Wales during 1944 to supplement No. 2 OTU. After only ten weeks at Narromine the new unit moved to nearby Parkes in September 1944. As of November 1944, No. 2 OTU focused on operational training for Kittyhawk pilots, while No. 8 OTU trained Boomerang and Spitfire pilots.

Paralleling No. 2 OTU's experience, mid-air collisions were responsible for all the fatalities No. 8 OTU suffered flying Boomerangs. On 7 December 1944 the Boomerangs flown by Pilot Officer Arthur Watkins (A46-32) and Flying Officer Harry Kidd (A46-36) collided during air-to-air combat training. Kidd opened his parachute and survived, but although Watkins escaped his burning aircraft his parachute did not open, and he was killed. Six months later on 29 May 1945 the Boomerangs flown by Pilot Officers Charles Weatherson (A46-82) and William Foster (A46-17) collided eleven kilometres north-east of Parkes airfield during a reconnaissance exercise. Both trainee pilots were killed.

While the Boomerang's RAAF service commenced with No. 2 OTU in late 1942, deliveries of Boomerangs to RAAF fighter squadrons finally commenced in 1943.

Interceptor Fighter Squadrons

With the prospect of significant numbers of Boomerangs being delivered in early 1943, the RAAF considered how to utilise them. It decided to establish three new "Interceptor Fighter" squadrons, Nos. 83, 84 and 85, which would absorb newly delivered Boomerangs and allow them to perform their designated role. Unfortunately, the squadrons' subsequent experiences showed that the Interceptor Fighter designation had been far too optimistic.

No. 83 Squadron was formed at Strathpine (now a Brisbane suburb) on 26 February 1943 under the command of Squadron Leader William Meehan. The squadron was controlled by

A profile of Boomerang A46-100 that was issued to No. 83 Squadron in June 1943. It received the fuselage code MH T. (Michael Claringbould)

the RAAF's Eastern Area Command, which was responsible for an enormous area of eastern Australia, including New South Wales and southern Queensland, and hundreds of kilometres of Australia's eastern seaboard.

The RAAF stated that the "purpose of the unit is the interception of unidentified and hostile aircraft, surface vessels and submarines". P-39 Airacobras were the only fighters available to No. 83 Squadron when it was established. Boomerangs started to join the squadron in April and by July they had completely replaced the P-39.

The squadron's Operational Record Book (ORB) describes an unvarying, unexciting and unrewarding list of second-line tasks carried out during the unit's tenure with Eastern Area Command, which lasted until the end of 1943. These tasks included the interception of unidentified (but always friendly) aircraft, cooperation with anti-aircraft and radar units, and training flights. The tedium of this regime may have contributed to the death of Flight Sergeant Edward Jones on 10 November 1943. Jones' Boomerang (A46-108) rolled onto its back and crashed at high speed near Caloundra in Queensland while performing unauthorised aerobatics. While these were prohibited, it is not surprising that some pilots indulged in aerobatics to escape the ennui associated with their predicable routine.

The squadron was transferred to the Northern Territory in the RAAF's North Western Area at the end of 1943. The Boomerangs operated from the RAAF stations at Milingimbi Island and Gove (now Nhulunbuy). These two bases were separated by 200 kilometres but the use of both extended the area the squadron could cover. The squadron's move to the Northern Territory transferred it to the RAAF's North Western Area Command. Fighter, bomber and reconnaissance units attached to this command were engaged in combat operations against the Japanese. However, No. 83 Squadron's responsibilities were:

> … monotonous and drab compared to the more glamorous roles played by the fighting & bombing squadrons, [but] 83 Sqdn's participation was no less exacting, comprising fighter cover for aircraft on sea reconnaissance, interception of all strange A/C [aircraft] reported in the vicinity, and protection for shipping convoys.

To its frustration:

> During the period spent in the NW area, the squadron flew several thousand hours, but not once was fortunate enough to be engaged in active combat with the enemy.

Under wartime conditions these routine duties could be deadly, as was demonstrated with shocking force in 1944. On 22 May the engine of the Boomerang flown by Pilot Officer Roy Ayre caught fire at low altitude. Ayre had no time to escape from his burning aircraft (A46-173), which crashed between Point Arrowsmith and Cape Shields. A week later Flight Lieutenant Nigel Pugh in A46-159 accompanied another Boomerang to relieve a RAAF Beaufort on a convoy protection mission. Pugh carried out a number of mock attack runs on the Beaufort but lost control of his aircraft on his last attacking pass and spun into the sea. No trace of Pugh was found.

After its tour in North Western Area, No. 83 Squadron and its Boomerangs returned to the backwater of Eastern Area Command, transferring to Camden, NSW, during August and September 1944. It made its final wartime move to Menangle, NSW, in February 1945.

Two more deadly incidents marred this otherwise routine period. On 20 March 1945 Warrant Officer Donald Wrightson (who was attached to No. 4 Aircraft Depot, based in Western Australia) was ferrying Boomerang A46-97 from Western Australia to No. 83 Squadron when he lost control of his aircraft in cloud and fatally crashed near Exeter, NSW.

The second of these incidents resulted in one of only two non-Australian Boomerang pilot fatalities. In mid-1945 the Royal Navy's No. 1843 Squadron, Fleet Air Arm, was based at Schofields (50 kilometres north of Menangle), where it was being trained for a planned deployment with the British Pacific Fleet. This training included exercises with No. 83 Squadron. As part of the normal familiarisation with each other's aircraft, Royal Navy pilots from No. 1843 Squadron flew to Menangle in their Grumman Corsairs on 30 May 1945 and then took off in Boomerangs. Royal Navy Sub-Lieutenant Kenneth Vogan took part in this seemingly routine exercise. However, shortly after take-off, Vogan's Boomerang (A46-76) dived steeply into the ground and crashed in a seemingly inexplicable accident.

No. 83 Squadron pilots with a Boomerang at Milingimbi, Northern Territory, in November 1943. (AWM)

No. 84 Squadron Boomerangs show the unit's "LB" fuselage code. (AWM)

No. 84 Squadron had the distinction of being the first frontline RAAF unit to operate the Boomerang. The unit was formally established on 5 February 1943 at Richmond, NSW, equipped with Australia's home-grown interceptor fighter. In April it deployed to Horn Island, located in the Torres Strait between Cape York Peninsula and New Guinea. The squadron was tasked with defending the Torres Strait region. This area included Merauke, a town on the southern coast of Dutch New Guinea which, unlike most of the Netherlands East Indies, had not been occupied by the Japanese.

During April the CAS confidently advised the War Cabinet that:

> The first squadron equipped with Boomerang aircraft is now at Horn Island and is ready for action … The Chief of Air Staff said that, although these aircraft were slower than Kittyhawks, they had a better rate of climb and were more manoeuvrable.

No. 84 commenced patrols over Merauke in May. It was directed to maintain a standing patrol of two aircraft over Merauke and also cover shipping within 32 kilometres of the town. These patrols became a key task for the unit during its brief period operating the Boomerang.

The squadron's Boomerangs attempted to engage Japanese aircraft three times. On 16 May Flying Officer Robert Johnstone (A46-51) and Sergeant Maurice Stammer (A46-60) attempted to intercept three Mitsubishi G4M Betty bombers while patrolling Merauke. This was the first encounter between Boomerangs and enemy aircraft. The Australian pilots closed to within 230 metres of the enemy bombers and opened fire. The guns of one Boomerang failed to fire, but the other directed 200 machine gun bullets and forty 20mm cannon rounds at the bombers. The Bettys returned fire with their turret and tail guns and escaped, undamaged, into clouds.

Over three months later on 30 August the squadron scrambled aircraft from Horn Island in response to an air raid warning, but the attempted interception was unsuccessful. Then on 9 September four Boomerangs took off from the recently completed Merauke airfield and rendezvoused with No. 86 Squadron Kittyhawks to intercept an attack on the airfield by 16 Bettys, which were escorted by 16 fighters. Ten of the 14 Kittyhawks were hampered by guns that partly or completely failed to fire, but the Kittyhawks were still able to report the destruction of three Japanese fighters. In an embarrassing contrast, the Boomerangs' participation was a complete failure. Unlike the Kittyhawks the Boomerangs were unable to reach the incoming raiders, which approached at 20,000 feet. The bombers attacked the airfield and destroyed grounded Boomerang A46-38.

No. 84 Squadron started to convert to Kittyhawks in September. The transition was complete by October, ending the Boomerang's brief and unremarkable career with the squadron.

On 12 February 1943 Squadron Leader Chris Daly was tasked with establishing No. 85 Squadron at Guildford, Western Australia, near the state capital Perth. The unit would provide fighter coverage in the RAAF's enormous Western Area Command, which included most of Western Australia.

The squadron was initially equipped with obsolete Brewster Buffalos but started to receive Boomerangs on 30 April. Guilford remained the squadron's main base, although detachments did operate from other locations. From Guilford the squadron carried out an unvarying routine of patrols, practice interceptions, gunnery exercises, cooperation with ground defences and night flying.

The dangers associated with such apparently routine flying were shown on 22 May 1943, when Sergeant Bruce Williams in A46-83 swung off the Guildford runway during an attempted take-off. The Boomerang burst into flames and Williams was badly burnt and died six days later. On 22 September 1943 Sergeant Boyd Wolf attempted to take-off after recovering from a bad landing but stalled and was killed when his aircraft (A46–41) crashed. Shortly afterwards, on 2 October 1943, A46-22 suffered a structural failure in the tailplane while being flown by Flight Sergeant Archie McDonald, who was killed in the subsequent crash. To complete a difficult initial period at Guildford, Flight Sergeant William Turnbull attempted to force land on 27 December 1943 after an apparent engine failure. The Boomerang (A46-68) overturned and caught fire, killing Turnbull. Fortunately, the squadron suffered no further fatalities operating the Boomerang at Guilford.

The burnt-out wreck of No. 85 Squadron Boomerang A46-83 which was destroyed after a take-off accident at Guildford on 22 May 1943. (AWM)

The basic facilities at the Exmouth Gulf airstrip codenamed Potshot in 1943. The photo has been taken from the control tower. (AWM)

On 1 May 1943 six of the Boomerangs that had just arrived at Guildford departed for the rudimentary airfield at Exmouth Gulf, codenamed Potshot, over 1,100 kilometres north of Perth. Potshot had been hastily constructed in late 1942 to protect a submarine base the United States Navy intended to establish nearby. The Boomerangs replaced No. 76 Squadron, which had operated Kittyhawks from Potshot after the base was completed in March 1943.

Exmouth Gulf is a remote location in what was, during the 1940s, a very remote state, separated by an enormous physical and psychological gap from Australia's more populous eastern seaboard. The pilots and ground crew battled to get the Boomerangs airborne under conditions where:

> … salt water entered fuel drums which had been unloaded from a barge and rolled onto the beach through water. Some spares, tools and equipment were lacking. A suitable pump for inflating undercarriage and tail wheel struts was never issued despite repeated requests.

Simply getting the Boomerangs into the air was difficult enough, but the detachment also had to be ready to fight the Japanese.

The Potshot-based Boomerangs attempted to engage Japanese aircraft on several occasions. On the night of 20/21 May Boomerangs piloted by Flight Lieutenant Roy Goon (A46-61) and Flying Officer Donald Goode (A46-58) were scrambled in response to radar reports of incoming aircraft. Although a bomb was dropped in Exmouth Gulf the Boomerangs did not contact the enemy aircraft. The next night, two Japanese aircraft dropped bombs in the gulf area and two Boomerangs again attempted an interception. Pilot Officer Lewellyn Wettenhall (A46-58) reported sighting the exhausts of a Japanese Betty bomber but had to break off the

A46-62 was one of the Boomerangs used at Potshot by No. 85 Squadron in 1943. (Michael Claringbould)

attempted interception when his fuel ran low.[1] Meanwhile, Flying Officer Malcolm Stevenson (A46-52) did not make any sightings. Months later, four Boomerangs scrambled on the night of 15/16 September following reports of incoming aircraft but did not make any sightings in the two hours they were airborne. This was the Potshot detachment's last attempt to intercept Japanese aircraft (despite its undistinguished WWII history, what began as a rudimentary landing field at Potshot is now RAAF Base Learmonth).

No. 85 Squadron also operated Boomerangs from the even more remote airfield at Derby, located 2,600 kilometres from Perth in Western Australia's Kimberley region. The first deployment took place in October 1943 in response to a perceived threat of a Japanese naval incursion into the Indian Ocean. The threat failed to materialise and the Boomerangs returned to Potshot in mid-October. Unfortunately, this operation was marred by a fatal crash on 3 October. Pilot Officer Brian Armstrong had been ordered to drop a message to the merchant vessel *Koomilya* at anchor in King Sound. Armstrong stood up in the cockpit to throw the message to *Koomilya* while flying over the ship at mast-head height. He lost control of his aircraft (A46-61), which rolled over and crashed into the sea.

Boomerangs also transferred to Derby in February 1944 and April-May 1944. They provided fighter protection for anchorages used by Catalina flying boats tasked with laying mines in Japanese controlled waters. In February RAAF Catalinas deployed to Cygnet Bay, 60 kilometres from Derby. The Catalinas mined Balikpapan harbour (on the east coast of Borneo) and its approaches. Boomerangs patrolled Cygnet Bay during the operation, after which they returned to Guilford. In April RAAF Catalinas deployed to Yampi Sound near Derby. No. 85 Squadron again transferred Boomerangs to Derby to protect the flying boat anchorage. The Catalinas mined harbours at Balikpapan, Sorong and Kaimana (the last two in the Dutch West New Guinea). A mine sank the Japanese destroyer *Amagiri* near Balikpapan on 23 April 1944. With the cessation of the Catalina operations, the Boomerangs left Derby by May 1944.

No. 85 began re-equipping with Spitfires during September 1944 and the last of its Boomerangs departed in January 1945, completing the squadron's operations with Australia's home-grown fighter.

---

1   The Japanese aircraft was in fact a No. 851 *Kokutai* Emily flying boat. Further details of these engagements are given in the Appendix.

The Boomerang's extensive service with the Interceptor Fighter squadrons yielded only a tiny number of fleeting contacts with Japanese forces. If the Boomerang's RAAF service had been limited to OTUs and Interceptor Fighter squadrons, it would never have had the chance to fight in the frontline of Australia's war against Japan. Ironically, it was when the Boomerang was employed in a role that was never considered during its frenetic development that it finally moved to the frontline and earned well-deserved battle honours.

Army Cooperation in New Guinea, Borneo and Bougainville

The Boomerang made its most important contribution to Australia's war effort in the army cooperation role in New Guinea, Borneo and Bougainville. This apparently unglamorous role included a number of tasks that provided vital assistance to Army units, such as very low altitude tactical reconnaissance, artillery spotting, photography and strafing opposing ground forces.

Fortunately for Australia's frontline soldiers, the RAAF had acknowledged the importance of army cooperation at the start of WWII. CAS Goble's proposed 1939 Home Defence Force included three army cooperation units among its nineteen squadrons. The RAAF's May 1942 Australian Air War Effort (effectively a blueprint for the massive wartime expansion of the RAAF) explicitly included cooperation with the field army as one of the air force's objectives. Perhaps one reason for the RAAF's clear-eyed recognition of the need to support the army was that Australia had no capacity to build the vast heavy bomber fleets constructed by Britain and the United States, which allowed key RAF and USAAF leaders to indulge their dreams of winning wars by strategic bombing alone.

The RAAF's commitment to working closely with the Australian Army was demonstrated when it established a School of Army Cooperation in January 1942, immediately after the start of war with Japan. Army officers and RAAF pilots attended the school and participated in each course. RAAF graduates of the course could then serve in the two dedicated army cooperation squadrons the RAAF established. These were Nos. 4 and 5 Squadrons: the unending demands of a world war prevented the formation of Goble's planned third squadron.

The RAAF's army cooperation role allowed both the Wirraway and Boomerang to operate in the frontlines, providing vital assistance to Australian soldiers fighting a ruthless enemy in a green tropical hell of malarial jungle, sweltering humidity, cloying mud, steep mountain ridges and seemingly endless close-quarter combat. The Boomerang enjoyed a significant combat career with the two RAAF army cooperation units, Nos. 4 and 5 Squadrons, which is detailed in the following two chapters.

CHAPTER 8
NO. 4 SQUADRON

In early 1943 the RAAF considered how to utilise the Boomerang after re-equipment of the Interceptor Fighter squadrons with the type had been completed. It decided that Boomerangs should be directed to its army cooperation squadrons. Pilots from No. 4 Squadron travelled from their New Guinea base to Australia in May 1943 and returned with Boomerangs in June. The Boomerang had finally gone to war.

The Boomerang would see service in a unit with superb capabilities in the army cooperation role. No. 4 Squadron was originally established in WWI as an AFC unit. The WWII iteration was formed at Richmond, New South Wales, in June 1940. After initially flying Hawker Demon biplane fighters it converted to Wirraways in September 1940. The unit was stationed in Canberra from October 1940 to May 1942, where it trained intensely in the army cooperation role, which included frequent exercises with Army units. After further movements to bases in New South Wales and Queensland, the squadron transferred to Berry Airfield near Port Moresby in New Guinea. On 20 November 1942 the squadron started to operate Wirraways over enemy occupied territory.

The Wirraway crews quickly gained experience in the types of missions they would soon be flying in Boomerangs. Tactical reconnaissance missions provided crucial assistance to the Diggers fighting a jungle campaign against an elusive enemy. These missions required dangerous flights across the Japanese lines at tree-top level, which allowed troop concentrations, fortifications, trenches and gun positions to be identified visually. Artillery reconnaissance, which involved observing where artillery rounds landed relative to the target and radioing corrections to the gunners, was exceptionally valuable. In addition, aerial photography was very useful considering the poorly mapped New Guinea of the 1940s, as it allowed accurate maps to be prepared to help plan ground and air operations.

Having gained operational experience since November 1942, No. 4 Squadron's pilots were exceptionally well placed to use the Boomerang in the army cooperation role when they started flying operations in the single seater in July 1943.

The Boomerang's introduction was quickly marred by an unfortunate "friendly fire" incident. The appearance and performance of the Boomerang differed from other Allied aircraft operating in New Guinea and the aircraft often came under ground fire from Allied units unused to seeing it. Flying Officers James Collier (A46-88) and John Utber (A46-89) departed from Wau on 5 July, having been tasked with carrying out a tactical reconnaissance at Salamaua. Cloud intervened and the pilots flew to the Nassau Bay area (30 kilometres south of Salamaua) to reconnoitre that location. US forces had recently landed at Nassau Bay and Collier sighted some of the invasion barges on the beach. Diving to inspect them, he was unaware that an earlier Japanese air attack had put the American forces on edge. American gunners opened fire on Collier with light anti-aircraft weapons. The Boomerang was hit, struck the water,

Two No. 4 Squadron pilots (wearing shirts) with ground crew and a Boomerang named Susanne at Nadzab on 5 October 1943. (AWM)

No. 4 Squadron pilots with their Boomerang at Nadzab in October 1943. The tailplane has recently been painted white to help with identification by friendly forces. (AWM)

skidded across the beach and crashed into the fringing vegetation. Collier was killed and the aircraft destroyed. He had the misfortune to be the first Boomerang pilot killed on operations.

Despite this setback No. 4 Squadron's Boomerangs quickly became a vital component of the Australian effort to push the Japanese back. After successfully defending Wau at the start of 1943, the Army pushed north towards nearby Japanese bases. Lae and Salamau were captured by September 1943, allowing the Australians to continue offensive operations to recapture the Huon Peninsula. To envelope the peninsula the 7th Division advanced generally north-west through the Markham and Ramu valleys and the 9th Division moved east towards Finschhafen.

No. 4 Squadron followed the Australian advance in the peripatetic way expected of an army cooperation unit. By mid-1943 the squadron was able to transfer many of its aircraft, including the newly operational Boomerangs, 240 kilometres north of the unit's original base near Port Moresby to an advanced airfield at Wau. In early September eight Boomerangs (and two Wirraways) moved further forward to Tsili Tsili, 80 kilometres west of Lae and Salamaua. By November, No. 4 Squadron had established two flights to support the Australian advance: A Flight supported the 7th Division from Gusap and B Flight supported the 9th Division from Nadzab.

Boomerangs quickly outnumbered Wirraways in No. 4 Squadron after their introduction in mid-1943. This situation prevailed until the end of the war, as detailed in the table below:

Boomerangs and Wirraways on the strength of No. 4 Squadron, June 1943 – November 1944

Month	Boomerangs Establishment/Strength/Available	Wirraways Establishment/Strength/Available
June 1943	-/4/2	18/12/9
July 1943	-/8/6	18/9/6
August 1943	9/12/10	9/9/7
September 1943	18/18/14	3/9/7
October 1943	18/17/14	3/8/6
November 1943	18/16/12	3/6/4
December 1943	18/16/13	3/5/4
January 1944	18/18/14	3/6/4
February 1944	(Missing)	(Missing)
March 1944	18/16/13	6/4/3
April 1944	18/14/10	6/3/2
May 1944	18/10/8	6/3/2
June 1944	18/12/9	6/3/2
July 1944	18/17/12	6/6/4
August 1944	18/16/11	6/6/5
September 1944	18/17/12	6/6/4
October 1944	18/18/13	6/6/4
November 1944	18/17/12	6/6/5

Records about Boomerangs on No. 4 Squadron's strength are not available after November 1944. However, the squadron's reconnaissance reports show that Boomerangs continued to carry out most of the unit's operational sorties for the remainder of the war.

In March 1944 B Flight moved from Nadzab to Cape Gloucester on the island of New Britain to support the US Sixth Army. Meanwhile A Flight continued to assist Australian forces as they pushed the Japanese along the north coast of New Guinea back towards, and then beyond, Madang and Alexishafen. After a mid-year hiatus when little operational flying was carried out by either Flight, B Flight returned to New Guinea in September 1944. No. 4 Squadron resumed operations in October 1944, flying from Tadji and Nabzab to support Australian efforts to destroy the disorganised remnants of the Japanese 18th Army in the Aitape – Wewak area on New Guinea's north coast. A small detachment briefly operated from Cape Hoskins in New Britain in early 1945 to support the 5th Australian Division.

In its final movement of the war, No. 4 Squadron departed from New Guinea in late March 1945 and relocated to Borneo to support the Australian campaign to liberate that island. No. 4 Squadron operated from airfields at Labuan (supporting the 9th Australian Division) and from Sepinggang, near Balikpapan (supporting the 7th Australian Division). Wirraways and Boomerangs commenced operations in June and July, respectively. Operations continued up to and after the Japanese surrender on 15 August 1945.

No. 4 Squadron's reconnaissance reports show that 56 Boomerangs assigned to the squadron carried out operational missions during the squadron's tour of duty (A46-46, -85, -88, -89, -92, -93, -95, -105, -106, -107, -109, -111, -112, -113, -115, -116, -117, -118, -119, -121, -132, -134, -136, -137, -138, -143, -145, -146, -148, -149, -151, -152, -156, -171, -172, -174, -179, -181, -183, -184, -187, -191, -193, -194, -195, -197, -198, -199, -209, -210, -211, -213, -217, -226, -229, -233).

Having briefly summarised the Boomerang's career with No. 4 squadron, we can look more closely at the range of missions the Boomerangs carried out to better understand the important contribution they made to the Army's fight against the Japanese.

A46-121 was a No. 4 Squadron Boomerang that carried the name Olga. It is seen at Balikpapan, Borneo, in 1945 by which time the white tails used in New Guinea had been removed. (AWM)

NARRATIVE REPORT-4(AC)SQUADRON,R.A.A.F.

LOCATION : GUSAP   5TH November 1943.

SORTIE No. 2.

ORDERS : TAC/R of BOGADHIM ROAD - KWATO-SUPPLY CAMP.

AEROPLANES   A46 - 134   F/Lt. OLERENSHAW (Recce)
AND PILOTS   A46 - 92   F/O. STALEY   (Cover)

                   up at 0850L
                   down at 1100L.

MAPS.   Photo-Maps - DAUMOINA - DAUMOINA SOUTH.

REPORTS

        The village of KWATO shows signs of recent occupation
although no movement was seen and there were no recent additions
to the number of huts.

        YAULA (125155) appears deserted and the huts are dilap-
idated. No movement was seen.

        OLD YAULA however appears inhabited and in good condition.

        At 123143 AN M/T road runs in a semi circle to 123141.
Cover aircraft A46.92 observed tracer coming from 4 or 5 camouflaged positions on this road.

        DAUMOINA village has been completely destroyed although
tracks leading through the village are still used.

        At DAUMOINA SOUTH at 140137 six fox-holes were seen, *whilst*
bridges 18 and 19 could be quite easily repaired.

        From KWATO to SUPPLY CAMP the road shows signs of heavy
use by M/T---the tread of the tyres were plainly visible.

A No. 4 Squadron tactical reconnaissance mission report for 5 November 1943 showing some detailed observations.

Tactical Reconnaissance

Tactical reconnaissance (TacR) missions were a crucial task for No. 4 Squadron's Boomerangs from their combat debut in mid-1943 until the end of war in the Pacific and beyond. The Boomerangs flew TacRs at low altitude, providing an astonishingly detailed picture of Japanese dispositions and movements. The Boomerangs carried out TacR at altitudes of not more than 200 feet, exposing them to rifle and machine gun fire. A contemporary account stated that during most missions the Boomerangs swooped down to 40 feet to carry out their missions, as this was the best altitude for observations.

During TacRs the Boomerangs typically operated in pairs. A reconnaissance aircraft operated at low altitude to carry out the assigned mission, while a second aircraft at higher altitude provided cover. This remained standard procedure throughout the war. The two aircraft could communicate directly via radio, while the reconnaissance aircraft could also contact high-level army headquarters. Japanese fighters were a serious threat in the initial months of the Boomerangs' operations and fighter escorts were provided during this period.

Effective cooperation with the army was critical to the tactical reconnaissance role. This was greatly assisted by an Army Air Liaison Section attached to the squadron, which ensured that tasking reflected the army's tactical needs. After each mission the Liaison Section compiled a detailed report of the pilot's observations, which provided superb intelligence to Australian ground commanders. The pilots could also send information directly to ground units. By dropping messages, information could be passed to battalion, brigade and division headquarters. On occasion, messages could also be dropped to smaller units if required. When telephone links had been established, pilots could telephone army units after landing and provide a summary of their observations.

The information from No. 4 Squadron's TacR missions greatly assisted the Diggers in their fight against an unwavering enemy. The Boomerangs supported the Army as it advanced north from Wau towards Lae and Salamaua, then carried out TacR missions to assist the 7th and 9th Divisions in capturing these two objectives in September 1943. The value of the information from TacRs increased as No. 4 Squadron's pilots learnt "tricks of the trade", such as flying at sea level to allow observation horizontally through the jungle when flying along coastal areas and detecting Japanese troops via the cooking fires they doused as the aircraft approached.

The Boomerangs' TacRs continued to assist the Army as it pursued defeated Japanese forces retreating from Lae and Salamaua. The Army's Official History notes the excellent level of cooperation between No. 4 Squadron and the ground forces achieved by late 1943. It highlights the speed, accuracy and efficiency with which the squadron's TacRs located Japanese targets in the jungle-clad, mountainous terrain then being contested.

The pilots were able to locate Japanese troop concentrations despite all attempts at camouflage by identifying individual foxholes, weapon pits and tracks worn in the jungle by the boots

A No. 4 Squadron Boomerang flies at typically low level over Australian troops in the Ramu Valley in January 1944. (AWM)

of Japanese soldiers. Vehicle movements were detected by the inevitable deterioration any vehicular traffic caused to the muddy tracks that passed as roads. This ability to identify Japanese troop concentrations and movements provided the Australian commanders with vital intelligence that was completely unavailable to their Japanese counterparts. This advantage was confirmed during the 7th Division's capture of Shaggy Ridge in January 1944, a decisive victory that allowed the Australians to continue their overland advance towards Madang on New Guinea's north coast. A Japanese map was captured during the battle, which showed that nearly all their main positions had been accurately located by No. 4 Squadron's TacRs despite every attempt to camouflage them.

Another demonstration of the Boomerang's ability to assist the Army took place in January 1944. The newly arrived and inexperienced 8th Brigade (part of the 5th Australian Division) was advancing west along the northern coastal strip of New Guinea as part of a second axis of advance towards Madang. Progress along the narrow coastal belt was hindered by numerous rivers and swamps. In a typical case of the assistance provided by No. 4 Squadron's Boomerangs, TacR missions on 26 January revealed that the Japanese had withdrawn from the Kwama River. This information allowed the Diggers to cross the river and continue their advance knowing they would be unopposed. The observations during this period were so thorough the pilots often reported the expressions on the faces of the retreating Japanese.

By August 1944 the 8th Brigade was largely occupied with locating and containing defeated Japanese remnants, including those trapped between the lower Ramu and lower Sepik rivers. No. 4 Squadron's Boomerangs were largely responsible for this task until the squadron's transfer to Borneo in 1945. The Boomerangs carried out TacRs at treetop altitudes to observe in meticulous detail villages, clearings, huts and even the gardens the Japanese were now using to try to avoid starvation. These TacRs reduced the Diggers' exposure to jungle fighting against a desperate opponent and undoubtedly minimised the loss of Australian lives.

The relentless scrutiny of Japanese troops continued even after Japan's surrender on 15 August 1945. No. 4 Squadron's Boomerangs continued their TacRs on Borneo during September, to locate Allied prisoners of war and monitor Japanese compliance with the surrender. It must have been with some satisfaction that Flying Officer Hansen reported on 5 September 1945 observations including:

> 1 single-engined 2 seater float plane (JAKE) … 4 SOUR faced Japs standing on Wings
>
> And:
>
> 7 MT [motor transport] 25 very unhappy Japs in Village .

No. 4 Squadron's Boomerang pilots paid a heavy price for the information their TacRs provided to the Army. Flying at tree-skimming altitudes exposed them to ground fire. In New Guinea (where most of No. 4 Squadron's missions were flown) an even greater danger was posed by low-level flying over dense jungle and jagged, sharply rising terrain in unpredictable tropical weather. The Boomerang's manoeuvrability and the skill of the pilots helped to counter these dangers, but potentially fatal risks were ever-present, as indicated by the list of Boomerang pilots from No. 4 Squadron killed on TacRs.

The skill and determination of No. 4 Squadron's Boomerang pilots in carrying out TacRs and the value of the information they provided were recognised by the Army, from senior officers to the Diggers engaged in close-quarter fighting. The Army Official History made this recognition clear when it reported the death of Flying Officer Bob Staley on 31 December 1943:

> On New Year's Eve occurred a loss which upset the fighting men of the 7th Division. One of the Boomerangs [A46-134] from No. 4 Squadron, piloted by Flying Officer Staley, was lost on reconnaissance near the 5800 Feature. These manoeuvrable tree-skimming aircraft, piloted by their valiant crews who seemed to know the tangled country so well, had appeared indestructible.

Artillery Reconnaissance

Artillery reconnaissance was a very important task for No. 4 Squadron's Boomerangs during their war service. It involved observing where artillery rounds landed and sending corrections to the gunners, helping them to accurately direct subsequent rounds onto the target. This procedure had been pioneered in WWI. Artillery reconnaissance was firmly established as an indispensable part of air support for ground forces during this conflict, despite air to ground communication being limited to one-way Morse Code transmissions from aircraft to the gunners.

No. 4 Squadron's pilots had already acquired valuable experience carrying out artillery reconnaissance in Wirraways before they commenced these missions in their new single-seat mount. As with TacRs, artillery reconnaissance was carried out by pairs of Boomerangs. Pilots were thoroughly briefed by a member of the Air Liaison Section before take-off. The briefing would include the number of targets, target coordinates and call signs. After take-off, radio

A46-204 seen at Nadzab on 1 August 1944 after a landing accident. The No. 4 Squadron "QE" fuselage code is clearly visible. (AWM)

No. 4 Squadron pilots and ground crew with a Boomerang named On the Job at Nadzab in October 1943.

communication would be established between the spotting aircraft and the artillery battery. When shooting commenced the initial ranging shots were often smoke shells, which clearly identified where the shells had landed and helped the pilot to provide accurate corrections to the artillery. The communication between pilot and artillery units was conducted with typical Australian informality, as typified by the exchange:

The gunner: "Is there a machine-gun post at —?"

The airman: "I'll have a look at the bastard"; and then, "The last shot was 100 yards over and 200 yards left—good shot though—I can see pots and pans flying all over the place."

While artillery reconnaissance missions may have lacked the dare-devil excitement (and danger) associated with low-level TacRs, they provided essential support to the Army. An unmistakable example is provided by the Army Official History's description of cooperation between No. 4 Squadron and the 9th Division during fighting around Sattelberg in November 1943:

During and after the war Japanese prisoners, high and low, claimed that the jungle artillery of the Australians was one of the most potent factors in their defeat, and bitterly compared its power with the comparatively ineffectual efforts of their own few artillery pieces which they hardly dared to use because of the Australian counter-battery fire and air support.

The artillery could not have been so effective without the excellent cooperation of the Boomerangs of No. 4 (Army Cooperation) Squadron of the RAAF which spotted for them. The audacious crews of these aircraft were very highly regarded by … the artillery no less than the infantry.

The Boomerangs' expert assistance acted as an effective force multiplier for the Australian artillery, which was unavailable to its Japanese counterparts. The technical advantage conferred by the Australian artillery/air combination provided a significant benefit to the hard-pressed Australian infantry. No. 4 Squadron's Boomerangs continued to provide this crucial support for the Army until hostilities ceased in August 1945.

Lead-in Strikes

The introduction of the Boomerang allowed No. 4 Squadron to carry out a new type of mission, the "lead-in strike", where the locations of targets were marked for attack by Allied fighters and bombers. In New Guinea, Allied aircraft attempting to strafe or bomb their targets were all too often confronted with a confusing, featureless carpet of jungle foliage that made accurate attacks impossible. The ability of the Boomerangs to confirm the locations of targets from low altitude and then mark them for attack enhanced the effectiveness of air attacks on Japanese ground forces.

Boomerang lead-in strikes commenced in September 1943, supporting Australian troops as they captured Lae and Salamaua. They continued at an increasing tempo as the Australian advance continued. The initial method was for Boomerangs, operating in the usual pairs, to mark targets by firing their cannons and machine guns at them using a high proportion of tracer ammunition. The Boomerangs also established radio communication with the attacking aircraft. The Boomerangs assisted RAAF and US aircraft attacking targets by strafing and both level and dive-bombing. Strike aircraft included P-39 Airacobra and P-40 Kittyhawk fighters (which often dropped bombs as well as strafed), B-25 Mitchell medium bombers and Vultee Vengeance dive-bombers. The Boomerangs often continued attacks after the strike aircraft departed by carrying out their own strafing runs.

While useful, marking targets by firing tracer ammunition in their direction was not optimal. Targets would ideally be marked by more precise and less transient methods. Rather than look to higher RAAF echelons for a solution, No. 4 Squadron devised a much-improved procedure at its own initiative. The squadron engineering staff modified a bomb rack compatible with the Boomerang to carry 13.6-kilogram (30-pound) smoke bombs. The Boomerangs were now able to mark targets directly with these bombs. The squadron proudly reported that:

> The Smoke Bomb Attachment … assisted immeasurably in army-air liaison with Boomerang aircraft. The use of smoke bombs in lead-in strikes is of inestimable advantage for use in thickly foliaged jungle terrain, diminishing margins for error and allowing accurate bombing of the directed target area. Another advantage of the smoke bomb procedure was that it allowed the Boomerangs to radio precise corrections to the strike aircraft during an attack.

Boomerangs were using smoke bombs by late February 1944. No. 4 Squadron's Boomerangs continued to carry out lead-in strikes for the rest of the squadron's war service.

Photo Reconnaissance

Photographic reconnaissance was one of the tasks carried out by No. 4 Squadron's Wirraways when they started active service against the Japanese in November 1942. The Wirraways

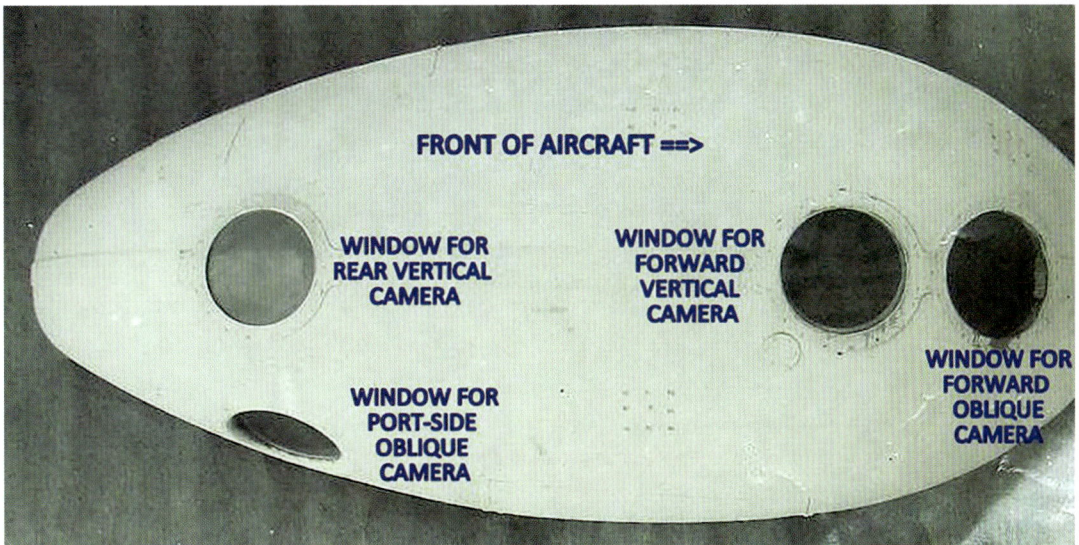

The underside of a Boomerang belly tank modified to accommodate two cameras. (NAA)

photographed enemy troop concentrations, gun emplacements and ammunition and supply dumps. Given the extremely rudimentary nature of existing maps of much of New Guinea, the squadron's crews also used their photographs as a navigational aid. As a versatile general purpose aircraft, the Wirraways could carry out photo reconnaissance missions without being modified. However, the Boomerangs were not equipped to carry out this type of mission when they commenced operations in mid-1943. Having been designed as an interceptor fighter in the febrile conditions of late 1941 and early 1942, adapting the aircraft for other tasks had not been a priority.

This deficiency quickly caused problems for No. 4 Squadron. When elements of the squadron were tasked with operating from Tsili Tsili to support the 9th Division in September 1943, photographic support for the division either required Wirraways to operate at excessive range or for Boomerangs to take photographs. This issue was addressed by a brilliant example of improvisation in the field. The squadron's engineering staff modified a standard Boomerang belly tank [drop tank] to accommodate two cameras. One camera could be installed in the forward part of the tank to take forward oblique or vertical photographs, while a second camera could be installed in the rear part to take vertical or port-side oblique photographs.

The openings for the cameras were uncovered in the initial version of the installation, but this allowed mud or dust to clog the camera lenses. The squadron's engineers then fitted Perspex (plastic) coverings to the openings. This improved design was introduced operationally in November 1943. Unfortunately, the Perspex covers distorted the camera images, so they were replaced by glass covers. Mud was often thrown onto the covers during take-off and the final step in the evolution of the belly tank camera was to tape cardboard discs to the covers, the discs being fitted with strings that were run to the cockpit. Pilots used the strings to detach the cardboard discs after take-off. This was the definitive belly tank camera installation, which allowed No. 4 Squadron's Boomerangs to provide the Army with effective photo reconnaissance support.

Displaying the instinctive controlling tendencies of large bureaucracies, the RAAF was wary of the squadron's local development of the belly tank camera when it learned of this initiative. RAAF Command demanded reports about the installation and considered whether it should be formally approved. The Command loftily decreed on 11 August 1944 that "installations of K17 and F24 cameras in Boomerang belly tanks not approved". However, this order quickly became redundant: all Boomerangs from A46-211 onwards were produced with a camera fitting in the rear fuselage that could accommodate an F24 camera. A46-211 was received by No. 4 Squadron on 20 August 1944.

Despite the Boomerang not being designed with photo reconnaissance in mind, the resourcefulness of No. 4 Squadron quickly allowed it to be used in this role. Another task was added to the range of army cooperation missions that the "Interceptor Fighter" was able to carry out.

Strafing

Boomerangs could attack ground targets with a powerful armament of two 20mm cannon and four 0.303-inch calibre machine guns. This provided a far more lethal punch than the Wirraway's two forward-firing 0.303-inch machine guns. No. 4 Squadron's Boomerangs strafed targets such as huts, vehicles, barges and troops during TacRs; after completing their marking duties during lead-in strikes; and in rare, dedicated strafing missions.

The enthusiasm of the pilots for strafing was noted in October 1943, when:

> It has been found that when pilots were briefed for Tac/R and to strafe opportunity targets the temptation is rather to look for straffing [sic] targets first and then to complete the Tac/R.

This report disapprovingly noted increased unserviceability due to hits from Japanese ground fire suffered while strafing. In response strafing was restricted to specific targets. However, No. 4 Squadron's reconnaissance reports make it clear that this restriction had little lasting impact on the pilots' keenness for directly attacking ground targets. Strafing continued to be combined with TacR missions throughout the squadron's wartime service.

Much of the Boomerangs' war was fought at low latitude, where pilots could easily see individual Japanese soldiers and even the expressions on their faces. This meant there was a far more close-up, personal aspect to many of their missions when strafing than was experienced by, for example, bomber crews. For better or worse, the outcomes of their strafing runs were often visible to the Boomerang pilots. An example is provided by a report of a mission in the Sepik River region of New Guinea on 2 February 1945, which noted:

> 1 Jap wearing shorts seen fishing in canoe on SEPIK RIVER. On strafing run JAP seen to have just reached bank. Cannon and MG fire knocked JAP onto edge of RIVER. On circuit Jap and blood seen in water.

RAAF armourers re-arm a Boomerang with 0.303-inch ammunition in 1944. (AWM)

No. 4 Squadron Boomerang A46-109, in which Flight Sergeant Alan Salter lost his life on 26 November 1943 when shot down by JAAF Oscars. (Michael Claringbould)

Paying the Price

No. 4 Squadron's Boomerang pilots had to fight against the Japanese, harsh terrain, treacherous weather and the dangers of low-level flying under wartime conditions. A number paid the ultimate sacrifice. The following chronological list of the squadron's Boomerang pilots killed on operations highlights both the dangers they faced and their determination to keep assisting their Army colleagues despite these dangers.

FLYING OFFICER JAMES COLLIER (A46-88), 5 July 1943: Flying Officer Collier's death is described on pages 63 and 64.

FLYING OFFICER THOMAS LAIDLAW (A46-112). On 6 September 1943 Flying Officer Thomas Laidlaw and Flying Officer Sydney Carter were tasked with carrying out a TacR in the Hopoi area, near Lae in New Guinea. The mission was intended to support the 9th Division, which had carried out an amphibious landing on 4 September with the objective of capturing Lae.

The two Boomerangs were attacked by Japanese fighters. Although Carter escaped the fighters and safely returned to base, Laidlaw failed to return. Laidlaw's aircraft was located in December 1948 and he was buried in Lae military cemetery.

FLYING OFFICER SYDNEY TRUMPER (A46-111). On 30 October 1943 Flying Officer Sydney Trumper and Flying Officer Dickson were carrying out a TacR to locate an enemy gun position between the Faria and Mindjim Rivers. Despite poor weather, with a cloud base only 100 feet above the area being reconnoitred, the Boomerangs descended to 50 to 100 feet above ground level to carry out the mission. Dickson could only see Trumper's aircraft intermittently. Eventually Dickson lost sight of him and could not locate any signs of Trumper's aircraft on searching the area. Australian forces subsequently captured the area and an Army patrol located the aircraft and Trumper in January 1944. The wreckage of the aircraft showed no signs of being struck by enemy fire. It appears that Trumper crashed in his determination to successfully complete the mission, despite the extremely dangerous weather conditions.

FLIGHT SERGEANT ALAN SALTER (A46-109) and **FLYING OFFICER HECTOR MUNRO** (A46-132): On 26 November 1943 these pilots were carrying out a TacR mission in the Sanga River area of New Guinea to support the 9th Division's advance from Finschhafen to Sattelberg. Enemy fighters remained a threat at this time and the Boomerangs were escorted by four P-39 Airacobras of the 41st Fighter Squadron, USAAF. However, the Airacobras were overwhelmed when the formation was attacked by a large number of Japanese Army Oscar fighters. The Boomerangs were separated from the P-39s and the Airacobra pilots last saw them being pursued by seven Oscars at 100 feet, near the mouth of the Sanga River. No trace of the two Boomerangs or their pilots was ever found.

FLYING OFFICER BOB (ERIC ROBERT) STALEY (A46-134): On 31 December 1943 Flying Officer Bob Staley was carrying out an artillery reconnaissance in the Mt Kubari area to support the 7th Division's assault on the razor-back feature known as Shaggy Ridge. Flying Officer Miller-Randle provided top cover. Miller-Randle lost sight of Staley during the reconnaissance but found a fierce fire on a hillside when he searched for his missing colleague. Staley did not return from the mission.

An Australian patrol was immediately dispatched to the area of the fire. The patrol located the wrecked Boomerang and Bob Staley's body on the 1 January 1944.

Flying Officer Lindsay Dann (A46-119) and **Pilot Officer Keith Hindmarsh** (A46-148): On 26 April 1944 Dann and Hindmarsh were tasked with carrying out a TacR in the Alexishafen area on New Guinea's north-east coast to support the 11th Australian Division. The two Boomerangs and pilots disappeared after departing from Gusap airfield at 0900. Searches failed to locate any trace of the two missing aircraft.

Postwar searches by Australian and US military personnel located the wreckage of the two Boomerangs and the remains of the pilots, and also established how the TacR had ended so disastrously. Both crashed aircraft were located in the Finisterre Ranges at an altitude of 9,000 feet. Local inhabitants stated they had seen the aircraft flying north along the Mungo-Curia valley. Dean and Hindmarsh spilt up when they approached a cloud-covered peak, one headed to the right of the peak, the other to the left. Unfortunately, both aircraft flew into the cloud-obscured terrain, about 1.6 kilometres apart. New Guinea's forbidding landscape and weather had claimed two more of No. 4 Squadron's Boomerangs and pilots .

Flying Officer Ronald Paxton (A46-116): At 0910 on 2 May 1944 Flying Officer Paxton took off from Gusap airfield to carry out a TacR north of Alexishafen to support the 11th Division. Paxton was accompanied by two other Boomerangs, flown by Flying Officer Beeck and Pilot Officer Oliver. Although the weather and visibility were good, Beeck and Oliver lost sight of Paxton's Boomerang after last seeing it at 500 feet about 160 kilometres north of Gusap. Paxton did not return. Searches at the time and post-war investigations found no trace of aircraft or pilot.

Flying Officer Lloyd Perry (A46-172): At 1330 on 2 May 1944 Flying Officer Perry took off from Nadzab for a test flight but did not return. Air and ground searches located the wreckage of the crashed Boomerang not far from the airstrip. Perry's body was nearby. His parachute was partly open, suggesting he had baled out of the aircraft too low for the parachute to fully open. The fuselage behind the engine was burnt out and it appeared the aircraft had caught fire in the air.

Flying Officer Kenneth Linklater (A46-85): Linklater was a member of the detachment based at Cape Gloucester on New Britain to support the US Army's 40th Division. At 0755 on 5 May 1944 Linklater and Flying Officers Rae and Lillie took off to carry out a TacR. At 0915 Rae and Lillie saw Linklater strike treetops during a strafing run, after a desperate attempt to pull up sharply and turn his Boomerang on its side failed to avoid tall trees. The wings and engine were torn off the smashed fuselage in an un-survivable crash. Linklater's unlikely target was four dogs suspected of being Japanese guard dogs.

With Linklater's death, No. 4 Squadron had lost five Boomerangs and their pilots in a space of ten days. This tragic period demonstrated just how dangerous the squadron's army cooperation role could be and the risks the Boomerang pilots faced every day.

PILOT OFFICER JOHN BUCKLAND (A46-137) and **FLYING OFFICER NORMAN OLIVER** (A46-138): On 27 June 1944 these pilots took off from Gusap airfield on a TacR to support the Australian 5th Division in the Hansa Bay area on New Guinea's north-east coast. Neither returned to base, being lost under circumstances that demonstrated the ever-present possibility that seemingly routine missions could end in sudden, unexpected death.

At 1315 members of the 5th Division at Hansa Bay saw Buckland's Boomerang descend to very low level over water as Buckland attempted to drop a message to their unit. The attempted low pass ended in disaster. A wing tip struck the water and the Boomerang burst into flames and sank almost immediately.

At the same time as Buckland's ill-fated message drop, Oliver attempted to make a wheels-up landing at Awar airfield, probably due to engine failure. Awar, near Hansa Bay, had been captured by the Army on 14 June and was available as an emergency landing field. Oliver could not have seen that a wrecked Japanese aircraft was hidden in grass on the runway. His Boomerang collided with the derelict aircraft and burst into flames. Unfortunately, Oliver was unable to escape.

SQUADRON LEADER CHRIS DALY (A46-149): Chris Daly had assumed command of No. 4 Squadron on 19 October 1944. On 17 November the new commander led a highly unusual mission for the squadron, an armed reconnaissance of the Little Ramu River area by seven Boomerangs to support the 8th Brigade. The aircraft lifted off from Nadzab at 0720 and by 0810 had arrived at the village of Angetji, the starting point for the reconnaissance. Daly swooped down to cross the river and strafe a target but misjudged his height after crossing the waterway and struck a tree as he pulled out of his dive. The Boomerang flipped onto its right-hand side and crashed into trees. The aircraft immediately burst into flames and burned fiercely for several minutes. The shocked survivors completed the mission and returned to Nadzab to report the loss of their commanding officer. This mission appeared to be the first and last time that No. 4 Squadron carried out an armed reconnaissance on this scale.

FLYING OFFICER RONALD SKUTHORP (A46-181): Flying Officers Skuthorp and Polkinghorne took off from Tadji at 0600 on 17 January 1945 to carry out a TacR to support the 6th Division. Skuthorp carried out the reconnaissance while Polkinghorne provided top cover. Reconnaissance of Japanese installations commenced at 0615 with targets also being strafed. From 0730 to 0745, Skuthorp investigated a ridge west of Wewak where he identified anti-aircraft guns, foxholes, machine gun posts and Japanese soldiers. At 0745, Skuthorp was circling at a mere 150 feet over anti-aircraft installations when his Boomerang was struck by machine gun fire. The aircraft lost height, crashed into trees and immediately exploded and burned. The squadron's commanding officer described Skuthorp's final mission in the following terms:

> The Reconnaissance Report … is indicative of the thoroughness of observation, of fearlessness, and of daring strafing of positions, the fire from which this pilot knew to be heavy and accurate. The Report is a classic of army co-operation reconnaissance effort, and within four hours of receipt by the Flight, 100 Squadron Beauforts were in the air,

followed later by 8 Squadron Beauforts … Flying Officer Skuthorp is missing, believed killed; but his final mission brought him retribution. His action was traditionally heroic: a tribute to the man, a tribute to his Squadron and his Country.

Fortunately, Skuthorp was the last Boomerang pilot No. 4 Squadron lost. His death meant that fifteen pilots from the unit were killed flying the Boomerang. This toll shows that army cooperation pilots operated in a high-risk environment, often flying at very low altitude in the face of hostile weather, unforgiving terrain and the enemy. Despite these dangers the squadron successfully carried out a wide range of army cooperation tasks with the Boomerang including TacR, lead-in strikes, artillery reconnaissance, photo reconnaissance and strafing.

CHAPTER 9
NO. 5 SQUADRON AND OTHER UNITS

The RAAF's No. 5 Squadron was the second unit to use the Boomerang in the army cooperation role. It was established at Laverton, Victoria, on 9 January 1941, equipped with Wirraways. The squadron moved to Toowoomba in Queensland in May 1942 and continued its training, including exercises with Army units, TacRs, photography and cooperation with anti-aircraft batteries. Further moves took the squadron to other locations in Queensland and further training with land forces units, including the 1st Australian Armoured Division, the US 40th Division and the 2nd Australian Corps. The unit accepted its first deliveries of Boomerangs in October 1943 while based at Mareeba in Far North Queensland. The type subsequently made up the majority of the squadron's aircraft, without completely replacing the Wirraway.

No. 5 Squadron finally moved to frontline service in November 1944 at Torokina on the island of Bougainville. Nominally part of New Guinea, Bougainville is geographically part of the Solomon Islands, where a bloody and distinct campaign had been fought since August 1942 by advancing American and New Zealand units. Bougainville's military importance was due to its location less than 400 kilometres southeast of the immense Japanese base at Rabaul on the island of New Britain.

Pilot Officer C Rasmussen from No. 5 Squadron in the cockpit of his Boomerang named Recce Robin at Mareeba in March 1944.

As the war turned against the Japanese, American planners looked to establish air bases on Bougainville that could be used in the campaign to neutralise Rabaul. American forces landed at Torokina in

Boomerangs and Wirraways of No. 5 Squadron lined up at Mareeba in mid-1944.

November 1943 and captured enough territory to construct airfields. The Americans then secured their perimeter and discontinued offensive operations. In November 1944 Australian Army units replaced the American troops on Bougainville, who were thus freed to participate in the American advance through the Philippines. Australia's political and military leadership decided to embark on an aggressive campaign to destroy the remaining Japanese forces on Bougainville instead of containing them, in a campaign often criticised as pointless. Whatever the merits of the campaign, No. 5 Squadron wholeheartedly supported the Diggers on the ground, flying its first mission to assist the Army from Torokina on 24 November 1944 and continuing operations until the Japanese surrender.

The types of missions flown by No. 5 Squadron's Boomerangs were similar to those flown by No. 4 Squadron in New Guinea. They included TacRs, artillery reconnaissance (including occasionally directing bombardments by Royal Australian Navy warships), lead-in strikes, photo reconnaissance and strafing (which was generally carried out against targets of opportunity during TacRs). One notable difference was that No. 5 Squadron's aircraft often operated singly, reflecting both a shortage of aircraft and pilots and the complete absence of Japanese fighters.[1]

The table below shows the aircraft strength of No. 5 Squadron, which generally had an establishment strength of eighteen Boomerangs and six Wirraways.

No. 5 Squadron strength of Boomerangs and Wirraways January 1945 – August 1945

Month	Boomerangs Establishment/Actual	Wirraways Establishment/Actual
January 1945	18/17	4/4
February 1945	18/18	4/4
March 1945	18/18	6/4
April 1945	18/19	6/4
May 1945	18/20	6/4
June 1945	18/21	6/4
July 1945	18/22	6/6
August 1945	18/23	6/8

The intense scrutiny of the Japanese by No. 5 Squadron identified defensive positions, the locations of artillery pieces, tracks used for transport purposes, barges critical to Japanese efforts to supply their troops and even gardens used to grow food for the isolated, starving garrison. After locating targets, the squadron often carried out lead-in strikes for Royal New Zealand Air Force (RNZAF) Corsair fighter bombers and RAAF Beaufort bombers. Most lead-in strikes were carried out by the Boomerangs, but No. 5 Squadron also used Wirraways to carry out these missions. Reprising No. 4 Squadron's experience, No. 5 Squadron quickly found that smoke bombs were far more effective for marking targets than strafing. The squadron reported that:

---

1   The Japanese Navy had withdrawn virtually all of its aircraft from Rabaul and surrounding areas in February 1944.

Experience has shown that indication of targets by dropping of smoke bombs is much more satisfactory than by strafing in jungle country. It is now customary for the Tac/R aircraft to observe the results of the bombing and from time to time, when necessary, drop further smoke bombs and/or correct the bombers … in a manner somewhat similar to the direction of an Arty [artillery] Shoot. Close co-operation and understanding with the RNZAF aircraft have resulted in numerous successful strikes, not the least of which have been several close support missions which have enabled our troops to advance and capture enemy positions with minimum casualties.

As this report shows, the squadron achieved excellent results when working closely with Bougainville-based RNZAF Corsair squadrons. The Corsairs operated as fighter bombers, typically dropping 1,000-pound bombs or naval depth charges. Both of these munitions could be equipped with a rod extending from the weapon's nose that caused it to explode slightly above ground level. This cleared the jungle without leaving craters to hamper an advance.

An excellent example of the coordination between the Boomerangs and the Corsairs was the destruction of Japanese tanks attempting a surprise armoured attack on Australian infantry. On 3 March 1945 a patrolling Corsair sighted Japanese tanks at Ruri Bay (in north-east Bougainville). The Corsair pilot radioed a request for No. 5 Squadron to dispatch a Boomerang to guide further Corsairs to the target. The squadron's commanding officer, Squadron Leader "Beau" Palmer, scrambled immediately and directed a flight of Corsairs to the target. Only one tank could be seen when the aircraft arrived, but Palmer located a second tank hidden in dense undergrowth. Both tanks were bombed and strafed, leaving one burning and the second damaged. A third was revealed when covering vegetation was shredded.

Next day No. 5 Squadron maintained a close watch on the area. The burnt-out tank was seen, but no sign of the other two was detected until Squadron Leader Palmer flew the day's final TacR as darkness approached. After several passes, Palmer located a damaged tank carefully camouflaged with branches and canvass. Responding to Palmer's report, Boomerangs flown by Squadron Leader Parry and Flying Officer Reynolds led four Corsairs to join Palmer. The Corsairs and Boomerangs bombed and strafed the hapless tank until darkness fell. No tanks reached the Australian lines, the attempt at a surprise armoured assault wilting under the joint ANZAC airstrikes.

Two No. 5 Squadron Boomerangs lead a strike by RNZAF Corsairs over Bougainville on 17 January 1945. (AWM)

One mission type where the Boomerangs did not shine with No. 5 Squadron was photo reconnaissance. Although Boomerangs did carry out this task, the fixed vertical mounting used by No. 5 Squadron did not provide the flexibility of the belly tank cameras used so successfully by No. 4 Squadron. No. 5 Squadron resorted to flying its Boomerangs tilted to the side to take oblique photographs. This practice was not particularly satisfactory and the squadron's Wirraways were primarily responsible for photographic duties.

In February 1945 the squadron sent a small detachment of four Boomerangs and a single Wirraway to Cape Hoskins on the island of New Britain. This replaced an element from No. 4 Squadron that was departing to rejoin its parent unit at Nadzab on the New Guinea mainland. In April the No. 5 Squadron detachment transferred to Tadji in New Guinea to support the Army after No. 4 Squadron had moved to Borneo. The detachment continued to operate from this location for the remainder of the conflict.

An idea of the variety of work carried out by No. 5 Squadron is provided by a breakdown of 221 missions flown in March 1945. More than half of these were Tac/Rs, although some of the specialised missions would have been flown by Wirraways:

Tac/R	128
Lead-in (bombers)	26
Arty/R	22
Photo/R	18
Strat/R	9
Strikes	7
Supply Dropping	5
Top Cover	4
Leaflet Dropping	2
Total	**221**

No. 5 Squadron's Tac/R Reconaissance Reports show that 32 of the squadron's Boomerangs carried out operational missions (A46-110, -128, -161, -163, -169, -175, -176, -178, -182, -186, -189, -190, -192, -196, -200, -201, -203, -211, -212, -214, -215, -216, -218, -220, -221, -222, -228, -232, -236, -237, -238, -239).

No. 5 Squadron operated in a generally less hostile environment than was experienced by No. 4 Squadron in New Guinea. It was never engaged by enemy fighters and the weather and terrain in Bougainville were typically less severe than those which cost No. 4 Squadron a number of Boomerangs and their pilots. Nevertheless, No. 5 Squadron did lose five Boomerang pilots, both while training in Queensland and during its operational deployment on Bougainville. These losses are listed in chronological order below:

PILOT OFFICER CAMPBELL MORRISON (A46-43): at 0830 on 14 January 1944 Pilot Officer Morrison took off from No. 5 Squadron's Mareeba airfield for an altitude test. During the flight the fuel in the port wing tank was exhausted and Morrison made the error of selecting

A Japanese tank burns after being attacked by No. 5 Squadron Boomerangs and RNZAF Corsairs on Bougainville on 3 March 1945. (AWM)

A camera is removed from a No. 5 Squadron Boomerang at Torokina in April 1945. (AWM)

A46-228 was a late build CA-19 Boomerang received by No. 5 Squadron on Bougainville in May 1945. (Michael Claringbould)

the unattached belly tank instead of the starboard wing tank. With the Boomerang running out of fuel Morrison attempted to land on a road. However, the aircraft struck a tree stump during the landing and overturned, killing the unfortunate pilot.

FLYING OFFICER WILLIAM THOMPSON (A46-48): Thompson took off from Mareeba at 0730 on 17 March 1944 to practise directing artillery fire. The engine cut out an hour into the flight. Thompson attempted a forced landing, but the aircraft crash landed and Thompson was severely injured, dying later the same day.

FLYING OFFICER ROBERT GRANGER (A46-192): On 5 November 1944 the departure of No. 5 Squadron's air echelon for active service at Torokina was approved. The high tempo of the squadron's training continued despite the imminent move and Flying Officer Granger took off from Mareeba at 1415 to carry out a TacR of local roads. While practising low altitude flying in hilly terrain Granger attempted a steep climbing turn to clear rising ground, but the port wing of his Boomerang struck a tree. The aircraft crashed and burned fiercely.

FLIGHT LIEUTENANT WALTER VERNON (A46-189): Flight Lieutenant Vernon took off from Torokina at 0732 on 11 January 1945 to lead-in an attack by 20 Corsairs from Nos. 21 and 24 Squadrons, RNZAF. Vernon used a smoke bomb to mark the target, which was attacked by No. 21 Squadron. The Corsair pilots then asked Vernon to drop a second smoke bomb to guide No. 24 Squadron's bombing runs. Vernon approached the target and dropped another smoke bomb, but the Boomerang appeared to stall at the moment of release. Disaster followed immediately as the Boomerang's port wing struck a tree and the aircraft crashed and exploded.

Vernon was the respected commander of the squadron's A Flight and his loss was keenly felt. By chance, a number of officers from No. 5 Squadron and No. 84 Wing (the higher RAAF formation that controlled the squadron) were carrying out a familiarisation flight in an Avro Anson at the time of Vernon's last flight and witnessed the explosion that marked his death.

PILOT OFFICER MORELAND OXLEY (A46-216): Oxley took off at 1530 on 6 February 1945 to lead-in two RNZAF Corsairs attacking a Japanese artillery position. Two Boomerangs were assigned to the mission and Oxley was accompanied by Pilot Officer Kidman. After a successful attack the Boomerangs headed back to Torokina across water at an altitude of 50 feet. Kidman saw Oxley's aircraft fly into the sea without any warning. The aircraft broke up and sank immediately. No trace of Oxley was found.

Other Units

Two other RAAF units are worthy of mention as Boomerang operators because they suffered fatalities due to Boomerang accidents.

The RAAF operated several Communication Units in New Guinea. No. 8 Communication Unit and No. 9 Communication Unit were formed in November 1943 by splitting No. 1 Rescue and Communication Squadron. Like their predecessor these units carried out a variety of tasks, including search and rescue, message dropping and transporting high value personnel and freight. The varied roster of aircraft operated by No. 8 Communication Unit included a number of Boomerangs. Unfortunately, two of the unit's pilots were killed while flying Boomerangs:

FLYING OFFICER HENRY KERSHAW (A46-94): On 24 December 1943 Kershaw was tasked with carrying a message from the RAAF's No. 9 Operational Group Advanced Headquarters at Vivigani on Goodenough Island and dropping it at the USAAF flying boat base at Samarai Island, some 160 kilometres to the south. Kershaw departed Vivigani at 0945 and dropped the message at Samarai at 1030 from an altitude of 100 feet. Kershaw had lowered the Boomerang's undercarriage to reduce speed during the message drop. As Kershaw attempted to gain altitude for his return flight a wing struck a tree. The Boomerang swung towards the sea, crashed into the water and sank, killing the pilot.

FLYING OFFICER DOUGLAS WILSON (A46-71): On 1 May 1944 three of the unit's pilots were authorised to practise formation flying. Wilson took off from Vivigani, accompanied by Flight Sergeants Gardner (A46-90) and Andrews (A46-95). After flying some 32 kilometres offshore the pilots looped their aircraft. Wilson was the last to carry out the manoeuvre but went into a spin at the top of the loop and crashed into the sea after failing to recover. The RAAF attributed the crash to poor technique.

One further pilot fatality was linked with the RAAF's operational use of the Boomerang in New Guinea, although the pilot was attached to No. 1 Aircraft Depot (a maintenance and logistics unit based at RAAF Laverton near Melbourne).

FLYING OFFICER ANTHONY BROOK: On 29 January 1944 Flying Officer Anthony Brook and Flight Lieutenant Bob Ashby were ferrying new Boomerangs A46-170 and A46-174, respectively, from Melbourne to New Guinea. This involved an arduous flight of more than 3,000 kilometres. Both Brook and Ashby were on exchange from the RAF to the RAAF. They took off from the remote, dusty RAAF station at Charleville in Queensland at 0740 to continue their ferry flight. Brook had gained 1,000 feet of altitude and was less than a kilometre from the airfield when he lost control of his Boomerang, which suddenly dived into the ground and crashed. The RAAF assessment of the crash indicated that Brook may have inadvertently locked the rudder of the Boomerang, rendering the aircraft uncontrollable.

An overhead aspect of A46-146 seen at Mareeba in 1944. (AWM)

Roy Goon (second from Left) at the Royal Victorian Aero Club. (Museum of Chinese Australian History)

Wackett trainers of No. 3 Elementary Flying Training School over Ballarat, Victoria, in February 1942. The instructor flying the third aircraft is Flight Lieutenant Roy Goon. (AWM)

CHAPTER 10
PROMINENT BOOMERANG PILOTS

A history of the Boomerang necessarily focuses on the aircraft, but we should never forget that the deployment of a warplane on active service generates an extraordinary range of human experiences, particularly for the pilots who flew the aircraft. However, accounts of Boomerang pilots are surprisingly sparse, and the following summaries are intended to address this gap. They represent pilots from the Interceptor Fighter squadrons based in Australia; No. 4 Squadron, which mainly operated in New Guinea; and No. 5 Squadron in Bougainville.

Roy Goon

Roy Goon's family name reflected his Chinese-Australian heritage. The first battle he faced in WWII was gaining entry to the RAAF in the face of barriers aimed at preventing non-Europeans from fully participating in Australian society.

Goon was born in 1913 in Ballarat, Victoria, and reportedly fell in love with flying during his first flight at the age of ten. He joined the Royal Victorian Aero Club (RVAC) in 1933, quickly acquiring experience in a wide range of types and becoming an instructor. His flying career took a dramatic turn in 1935 when he joined the air component of Chiang Kai-Shek's Republic of China forces as an instructor: Chiang's forces were fighting against Japan in the Second Sino-Japanese War. Goon rejoined the RVAC on his return to Australia.

Goon applied to join the RAAF as soon as WWII broke out. However, he was rejected twice because the Commonwealth Defence Act then barred applicants "not substantially of European origin or descent" from serving in Australia's armed services. Demonstrating his well-known initiative, Goon secured the support of Minister for Air James Fairbairn and gained his commission when he reapplied in mid-1940. He was the first Chinese-Australian to do so. He had already flown more than 2,700 hours on aircraft including the de Havilland Moth and Tiger Moth, Avro Cadet, Miles Falcon and Whitney Straight, Hawker Osprey, Waco C and D types, Stinson Reliant and the Boeing P-26 Peashooter, as well as qualifying as an aircraft engineer.

Goon's initial postings were to training units, where his students benefitted from his broad experience, renowned flying ability and calm, unflappable demeanour. He was promoted to flying officer in August 1940 and flight lieutenant in January 1942.

Goon's career took a dramatic turn when Japan brought the war to Australia's doorstep in early 1942 and the demand for pilots in frontline squadrons increased dramatically. In May 1942 Goon was posted to No. 34 Squadron at Darwin, which carried out transport flights across northern Australia. He was transferred to No. 24 Squadron at Townsville in July 1942. This unit carried out an essential but unexciting range of "general purpose" duties, including anti-submarine patrols, confirming the identity of friendly aircraft, cooperation with ground-based anti-aircraft defences and training.

Goon's next transfer took him to No. 2 OTU at Mildura in January 1943, where he was introduced to the Boomerang. He joined No. 83 Squadron at Strathpine, Queensland, in early April 1943 when the unit was being equipped with Boomerangs. His stay at No. 83 Squadron was short, as he was one of eleven pilots who flew Boomerangs thousands of kilometres across Australia to transfer them to No. 85 Squadron at Guilford, Western Australia, at the end of the April 1943. After completing this marathon transfer flight on 30 April, Goon and five other pilots flew another 1,100 kilometres to the Potshot satellite airfield at Exmouth Gulf the next day. On 20 May Goon was one of two Boomerang pilots scrambled from Potshot at 2240 in response to a radar report of unidentified incoming aircraft. Unfortunately, despite directions from ground control, neither pilot sighted the unidentified aircraft. This proved to be the closest that Goon ever got to engaging enemy aircraft. He continued to fly the Boomerang with No. 85 Squadron throughout the rest of 1943. His dedication to duty and leadership skills were rewarded with promotion to the rank of squadron leader on 1 August.

Goon attended the RAAF's Staff School in November and December 1943, and was then appointed to command No. 83 Squadron. This squadron transferred from Strathpine in Queensland to Gove in the Northern Territory over the late 1943/early 1944 period. Goon arrived at Gove in January 1944 to assume his new command. The squadron's operations during its deployment in the Northern Territory were unvaried and maintaining the unit's enthusiasm during this period cannot have been easy. Goon rose to this challenge and was Mentioned in Despatches for his leadership. According to his citation for the award:

> Squadron Leader Goon was posted to command No. 83 I/F Squadron which moved to North-Western Area in January 1944.

> No. 83 was responsible for the protection of convoys and the shipping route from HORN ISLAND to DARWIN and the base at GOVE. These shipping missions were particularly arduous. They necessitated long flights over the sea in single-engine aircraft in all weathers.

> Squadron Leader Goon displayed conspicuous leadership and devotion to duty and was at all times an inspiration to all personnel under his command.

Goon's time with the Boomerang finished in February 1945, when the RAAF posted him to command the Queensland-based No. 111 Fighter Control Unit (FCU). No. 111 FCU deployed to Morotai in the Netherlands East Indies in May 1945 and to Labuan, Borneo, in June 1945 to direct Allied fighters. Goon left the RAAF in October 1945.

Simply joining the RAAF required initiative and persistence from Roy Goon. He flew Boomerangs for two years during his distinguished wartime career, rising from his introduction to the type at No. 2 OTU to command of No. 83 Squadron. While he did not engage enemy aircraft or attack enemy ground forces, he played his role with determination and wholehearted commitment, doing all he could to support Australia's war effort. This must have been an attitude shared by many of his fellow Boomerang pilots in the Interceptor Fighter squadrons.

Bob Staley

In March 1941 English-born Eric Robert "Bob" Staley travelled 400 kilometres from his farm in the tiny hamlet of Natya in north-west Victoria to Melbourne to enrol in the RAAF. Staley had no prior military or aviation experience but qualified as a pilot in September 1941 after completing his initial ground training, Elementary Flight Training (flying Tiger Moths) and

Rosa Staley, Eric Staley and Bob Staley in 1941. (AWM)

Service Flying Training (flying Wirraways). Staley was posted to No. 5 Squadron in December 1941, which operated Wirraways in the army cooperation role.

Staley received extensive training in a range of army cooperation duties during his lengthy posting with No. 5 Squadron. Soon after being transferred to the unit, he took part in the first course run by the RAAF's School of Army Cooperation. Staley returned to his squadron in February 1942 and flew an extensive range of training missions, including TacRs, artillery spotting, practice bombing and strafing, photography and cooperation with anti-aircraft defences. By the time he was posted to operational flying with No. 4 Squadron in August 1943, Staley had more than two years of flying experience and was thoroughly trained in army cooperation tasks.

Staley was one of five replacement pilots who arrived at No. 4 Squadron's base at Berry airfield near Port Moresby, New Guinea, on 1 August 1943. He could finally put into practice the long months and years of training that preceded his arrival at the frontline. At the start of September, Staley joined a detachment that transferred eight Boomerangs and two Wirraways to Tsili Tsili to support the Army's successful operation to retake Lae and Salamaua from the Japanese.

In mid-September Staley carried out an extremely dangerous night flight to transport captured Japanese plans from Tsili Tsili across the towering Owen Stanley Mountains to Army headquarters in Port Moresby. He flew a Wirraway on this occasion. Staley was Mentioned in Despatches for this feat and was also awarded the "Nabzab Cross", improvised by his squadron mates from the lid of a food container.

By November No. 4 Squadron had established a detachment at Gusap to support the 7th Division. Staley flew Boomerangs from this base, carrying out the typical range of TacRs, artillery reconnaissance, lead-in strikes and strafing associated with No. 4 Squadron.

The "Nadzab Cross" improvised medal presented to Bob Staley by members of No. 4 Squadron in September 1943. (AWM)

Staley continued to fly from Gusap in December, increasing his tally of Boomerang missions. He had a particularly busy and productive day on 9 December. The day started with a TacR that departed Gusap at 0640. Staley carried out the TacR in A46-92, while Flying Officer Guyot provided cover in A46-134. Ignoring ground fire from several positions, Staley identified two ridgelines dotted with foxholes and estimated that at least 300 Japanese soldiers were present in the area. He dropped messages with this vital information to the Australian 2/6 Cavalry Commando Squadron, which was desperately seeking information about Japanese dispositions. The importance of this mission is indicated by it being specifically mentioned in the Army Official History. Staley and Guyot took off for their second mission of the day at 1220, leading a bombing and strafing attack by eighteen P-40s.

Four days later, on 13 December, Staley (A46-119) flew cover when Flying Officer Miller-Randle (A46-118) carried out a TacR of Japanese troops apparently attempting to outflank the Australian 25th Brigade. The intelligence from this TacR was integrated with other information, leading to a relocation of the brigade's units to avoid being outflanked. This mission is also highlighted in the Army Official History. Staley's routine of flying TacRs in Boomerangs was varied later in the month, when he piloted Wirraway A20-619 on 21 December on a mission to drop supplies to isolated units of the 25th Brigade. The next day Staley, in Boomerang A46-118, located a crashed American B-25 Mitchell medium bomber while being covered by Flying Officer Eller (A46-143). A ground patrol was directed to the crash site. Staley (A46-118) located the crew the next day and dropped a message advising them to remain in place until the patrol reached them.

By this time Staley had distinguished himself as an outstanding pilot. This was reinforced on 27 December when he was selected to fly the commander of the 7th Division, Major General Alan Vasey, over the battlefield in a Piper Cub as Vasey observed his division's progress dislodging the Japanese from Shaggy Ridge.

After starting operations with No. 4 Squadron, Bob Staley quickly showed that he was a brave, skilful and determined pilot. He carried out the squadron's diverse range of army cooperation missions, mostly in the Boomerang but also using the Wirraway. It seemed likely that further honours and achievements lay ahead of this outstanding officer. Before taking off in Boomerang A46-134 for a TacR on 31 December 1943 Staley wrote to his wife, saying "today is the last day of 1943. I wonder what will happen next year". However, as described on pages 77 and 78, this day was the last day of both 1943 and Bob Staley's life. He was mourned by his colleagues in his squadron and by his wife and son.

Beaufort "Beau" Palmer

Queenslander Beaufort Mosman Hunter "Beau" Palmer was born in Brisbane in 1919. A keen sportsman, he represented his school at rowing, athletics, swimming and rugby. After leaving school he worked as a shipping clerk, but this uneventful occupation did not appeal to Beau and he applied for an RAAF cadetship. He joined the RAAF as an Air Cadet at No. 1 Flying Training School at Point Cook on 16 January 1939 (the more structured wartime training system had not been introduced at that time). Palmer successfully completed his training and was commissioned as a pilot officer in October 1939.

Palmer's first posting after training was to No. 23 Squadron in his home state of Queensland, where he flew Wirraways. His career with this squadron featured an incident on 11 April 1941 when he force-landed Wirraway A20-25 in a Brisbane Street. Palmer had suffered an engine failure while performing aerobatics near the home of his future wife. A court of inquiry was convened, and an RAAF report of the incident carried the unusual note "seen by Minister". Fortunately, Palmer's career survived, and he notched up another highlight when he was the first pilot to land at the new RAAF base at Amberley in July 1940 (before the official opening) to deliver a consignment of alcohol to the officers' club.

Beau Palmer during an inspection of RAAF Archerfield by Prime Minister Robert Menzies in August 1941.

Palmer was posted to No. 24 Squadron, which formed at Amberley in July 1940 operating Wirraways. He carried out instructional duties at Service Flying Training Schools from October 1940 to June 1942 and was promoted to flight lieutenant in July 1941. In a sign of things to come, Palmer was posted to the School of Army Cooperation in June 1942 as Assistant Chief Instructor. He remained at the school until April 1943, when he was transferred for duty as a fighter controller with Queensland-based fighter control units. He was promoted to squadron leader in October 1942.

After many months of service far from the frontlines, the possibility of combat duty appealed to the high-spirited Palmer and he secured a second posting to the School of Army Cooperation in May 1943, this time as a trainee. After completing the course, he received further instruction at Operational Training Units and the RAAF Staff School. Finally, in November 1944, Palmer was posted to No. 5 Squadron at Bougainville for active service.

After almost six years of service with the RAAF and years of non-combat postings, Palmer must have been thrilled when he flew his first combat sortie on 7 December 1944 in Boomerang A46-221. As if making up for lost time, Palmer energetically carried out all the squadron's mission types, including TacRs, artillery reconnaissance, photography, lead-in strikes and strafing. He flew both Boomerangs and Wirraways, logging more hours in the compact single-seater. Palmer made the further transition from combat pilot to combat leader when he was appointed as No. 5 Squadron's commanding officer on 30 January 1945.

Palmer's notable role in the defeat of an attempted Japanese armoured attack has already been described. Another of his favoured missions was attacking the small watercraft the Japanese relied on for transport in Bougainville. On 5 February 1945 Palmer (A46-182) and Flight Lieutenant Saunders (A46-201) followed up a sighting of a barge made by No. 5 Squadron earlier that day on the northwest coast of Bougainville. They located the barge and strafed it with cannon and machine gun fire, noting many strikes. The vessel was claimed as destroyed. Taking off before first light on 9 February, Palmer (A46-200) and Flying Officer Witford (A46-214) orbited near the mouth of the Puriata River in southwest Bougainville in an attempt to locate and attack barges at first light. On this occasion no vessels were seen. Another search of the same area next day by Palmer in A46-200 was similarly unsuccessful.

Palmer had more luck on 21 March, flying to Bougainville's northeast coast in A46-212 to follow up a sighting of a barge made by his squadron earlier in the day. Palmer led-in two New Zealand Corsairs that bombed the barge. After the bombing Palmer and the Corsairs repeatedly strafed the barge which, unsurprisingly, was left awash and drifting. A further barge was then spotted and strafed by the three aircraft, with cannon and machine gun strikes and an oil slick being seen. On 23 April Palmer (A46-200) spotted a 20 feet motor launch hidden under trees during a TacR of Bougainville's northeast coast. He strafed the launch with 100 20mm cannon rounds and 2,000 machine gun rounds, leaving the vessel filled with water and claimed as sunk.

Palmer continued his war against the Japanese watercraft on 11 May. During a very busy day, he spotted and strafed a barge and small boat on the northwest coast during a TacR in A46-237. He returned to bomb these vessels in Wirraway A20-502, accompanied by Flying Officer Reynolds (A20-588). The bombing revealed three more barges, a small boat and two canoes. After completing the bombing sortie, Palmer returned in Boomerang A46-169 to lead-in an attack by eight Corsairs. The day's final tally was one barge and a small boat either sunk or damaged, one barge damaged and the canoes accurately strafed. Palmer's missions against Japanese vessels were only one of the many types of missions he flew, but they serve to illustrate the energy, initiative and commitment he brought to his role as commanding officer. However, late in May his strikes against Japanese vessels ended abruptly, along with his tour of duty.

Palmer was hunting pigeons instead of the enemy on 29 May. This diversion from his normal responsibilities was shattered when he stood on an old American anti-personnel mine. The resulting explosion almost severed his right foot and shrapnel entered his left foot and buttocks. After immediate first aid by other members of the shooting party, the squadron medical officer arrived and supervised Palmer's transfer to hospital. Palmer remained conscious during this ordeal and when asked if anything could be done after the explosion replied "yes, you can go over there and get my bloody foot". He was repatriated to Australia in June 1945 and discharged from the RAAF on medical grounds in April 1947.

Palmer's service was recognised by the award of a Distinguished Flying Cross (DFC). His citation stated that:

> … Squadron Leader PALMER has, at all times, displayed zeal, initiative and courage in

excess of the normal call of duty. At all times, he has undertaken the more dangerous or difficult sorties, and has always displayed outstanding leadership in the air.

Important sightings of tanks, vehicles, barges, dumps and defended localities by Squadron Leader PALMER have materially aided ARMY Forces in conducting the Bougainville campaign.

He has … carried out 108 sorties against the enemy.

The award of the DFC formally recognised Palmer's transition from brash junior pilot to responsible, energetic commanding officer. Flying the Boomerang in combat was an important part of this transition.

CHAPTER 11
ASSESSING THE BOOMERANG

The Boomerang would scarcely have seen combat if its wartime use had been confined to Operational Training Units and the "Interceptor Fighter" squadrons. Fortunately, the Boomerang was subsequently used in the army cooperation role, where it proved highly successful. The Army Official History highlights the effectiveness of the Boomerang in this role and records the gratitude of the Diggers for its support.

Much of its success as an army cooperation aircraft was due to the Boomerang's inherent virtues. The Boomerang was compact, comparatively fast (particularly compared to the Wirraway) and highly manoeuvrable. These qualities made it well suited to carrying out low-level reconnaissance missions in New Guinea's unforgiving terrain, while reducing its vulnerability to ground fire. The aircraft was rugged and coped well with operating from rudimentary forward airfields, achieving excellent serviceability levels. The Twin Wasp engine was very reliable and proved extremely resistant to harsh operating conditions, while the Boomerang's heavy armament provided a useful punch for attacking ground targets. The Boomerang proved eminently suited to the army cooperation role, where it found its true niche as a combat aircraft.

Despite these admirable qualities, going to war in the Boomerang and training to go to war were dangerous. A total of 38 RAAF pilots were killed operating the Boomerang in WWII. They flew with the following units:

No. 2 OTU	2
No. 8 OTU	3
No. 83 Squadron	5
No. 85 Squadron	5
No. 4 Squadron	15
No. 5 Squadron	5
No. 8 Communication Unit	2
No. 1 Aircraft Depot	1

As well as these fatalities, many other Boomerang pilots must have suffered physical or psychological scars that affected them long after the war ceased. While acknowledging the sacrifices made by these killed and maimed pilots, we now turn to an assessment of the Boomerang's wartime service.

Overall, was the Boomerang a "success" or "failure", despite the risks of reducing a complex judgement to a single word? An assessment based on the War Cabinet's reasons for ordering the Boomerang gives one answer, but a different approach gives a different answer.

The War Cabinet's February 1942 decision to order 100 Boomerangs rested on two arguments:

- it would provide a stop-gap fighter in case imports of more capable fighters did not reach Australia in time to respond to the Japanese threat.

- Boomerang production would utilise CAC staff who would otherwise run out of work as Wirraway production wound down, pending their transfer to production of the CA-4 bomber.

Subsequent decisions were heavily based on avoiding disruption to CAC's production regime, particularly its workforce. The War Cabinet's October 1942 decision to acquire 100 more Boomerangs was prompted by CAC's advice that further orders were needed to provide work for the staff producing the aircraft until manufacture of the CA-4 commenced. In July 1943 the War Cabinet reluctantly agreed to order an additional 50 Boomerangs to help maintain CAC's manufacturing capacity until the company started to produce the North American P-51 Mustang fighter.

The Boomerang as a Stop-Gap Fighter

The Boomerang's performance was completely inadequate for the air-to-air combat role. Wing Commander Peter Jeffrey's summary, quoted in Chapter 7, bluntly details the Boomerang's flaws as a fighter aircraft. There are also the results of the Boomerang's actual engagements with Japanese aircraft summarised in the Appendix. Although a small sample, these results suggest large-scale use of the Boomerang in its nominal "Interceptor Fighter" role could have been disastrous.

As well as the Boomerang's lacklustre performance as a dedicated fighter, the excessive time taken to deliver the aircraft is critically important. Success as a stop-gap fighter required delivery of Boomerangs in meaningful numbers in the period when an active Japanese aerial threat to Australia existed but before significant numbers of imported (i.e. American) fighters reached Australia.

In reality, it took CAC until mid-1943 to deliver 100 Boomerangs to the RAAF. Japanese forces in the South West Pacific were on the defensive by then and the aerial threat to Australia had greatly reduced. As a comparison the RAAF had received 293 capable and versatile P-40 Kittyhawk fighters from the US by June 1943. Initial deliveries of Kittyhawks were sufficient to allow the RAAF to establish three new fighter squadrons (Nos. 75, 76 and 77) in March 1942. These squadrons all saw active service in 1942. In contrast, Boomerang deliveries were far too slow to have made a useful contribution to Australia's aerial defence.

The Boomerang's inferior performance and sluggish delivery to the RAAF clearly show that it failed as a stop-gap fighter.

Utilising CAC's Fishermans Bend Workforce

As well as providing fighters for the RAAF, the February 1942 and October 1942 decisions to acquire Boomerangs were intended to provide work for CAC staff until production of the

CA-4 bomber started. The July 1943 decision to order 50 more Boomerangs was designed to preserve CAC's manufacturing operations, including its workforce, until production of the P-51 Mustang fighter commenced.

However, the tangible results of both the CA-4 and Mustang projects were dramatically different from the optimistic forecasts presented to the War Cabinet. The CA-4 Woomera, designed by CAC under Lawrence Wackett's guidance, was a twin-engine, multi-purpose bomber/reconnaissance aircraft. The highly ambitious design included remote-controlled machine guns in the rear of the engine nacelles to fend off attacking fighters and liquid-tight wing cavities to serve as fuel tanks. The complexities associated with the aircraft were never satisfactorily solved. The project suffered a huge setback when the prototype exploded in mid-air in January 1943, killing two of the three crew. The CA-4 never entered production despite absorbing much of CAC's engineering resources during its protracted development. The Mustang project was similarly disappointing. Intended to provide the RAAF with a locally produced world-class fighter, CAC built the aircraft from a mix of imported and locally manufactured components. However, instead of the projected mid-1944 date, the first CAC-produced Mustang flew on 29 April 1945. It was delivered to the RAAF in June 1945. With only seventeen aircraft delivered to the RAAF before Japan's surrender the CAC Mustang arrived far too late to make any contribution to Japan's defeat.

Judged by the War Cabinet's reasons, the Boomerang was a clear-cut failure. It was unsatisfactory as a stop-gap fighter and the CAC resources kept in place to build the Boomerang were subsequently used to produce aircraft of negligible value to Australia's war effort.

Despite this seemingly damning conclusion there are other ways of assessing the Boomerang that produce a more favourable verdict.

An Assessment Based on Benefits and Costs

The Boomerang does not have to be judged solely by the criteria set by the War Cabinet. Wars are inherently unpredictable events, where fortunes can change surprisingly quickly, and decisions are made using limited information. Another way to assess the success or failure of the Boomerang is to compare the benefits the Boomerang provided to Australia's war effort with the costs of producing it.

The Boomerang was issued to the Australian-based Interceptor Fighter Squadrons and was the major type used by No. 83 Squadron (based in eastern and northern Australia) and No. 85 Squadron (based in Western Australia). These units carried out a range of unexciting home-front tasks, such as confirming the identities of incoming aircraft, cooperation with ground-based defences and convoy patrols, but these routine duties were unavoidable in a sparsely populated island continent with an enormous coastline. The Boomerangs provided valuable service by carrying out these missions.

While the Boomerangs used by the Interceptor Fighter squadrons performed second-line tasks, the Boomerangs assigned to Nos. 4 and 5 Squadrons provided direct and effective support for the Army. The valuable nature of this contribution was highlighted in Chapters 8 and 9 and is confirmed by the Army Official History. By war's end, 88 Boomerangs attached to the two squadrons had participated in missions against enemy forces, showing that a significant part of the Boomerang's production run of 250 saw active service in the army cooperation role.

While recognising the positive contributions the Boomerang made to Australia's war effort, we also have to consider the resources used to produce it.

With Australia mobilising for total war and all available industrial capacity being directed to military production, the real "cost" of producing the Boomerang was the use of CAC resources to produce 250 aircraft at Fishermans Bend.

As former RAAF Air Vice-Marshal Brian Weston points out, CAC's resources were shamefully under-utilised during WWII.[1] This unhappy situation reflected a lack of timely orders to follow the Wirraway and the establishment of a completely new organisation to produce the Beaufort bomber instead of ordering it from CAC. The neglect of CAC's capacity is confirmed by the argument put to the War Cabinet that Boomerang production was needed to utilise much of the Fishermens Bend workforce, which otherwise would have lacked productive work and have to be dispersed. Boomerang production did not displace other projects that would have made a larger contribution to Australia's war effort.

The overall picture is that the Boomerang performed valuable frontline and second line tasks and was produced using available capacity at CAC. Considered in these terms, the Boomerang project can be considered an effective use of resources and hence a "success".

The Boomerang's most important contribution to Australia's war effort was the assistance it provided the Army in New Guinea, Bougainville and Borneo. Its success in this role was amplified by a number of factors, including the RAAF's commitment to the army cooperation role, the dedication and bravery of the pilots and the ingenuity of ground staff, whose field modifications provided additional roles for the Boomerang. This combination of professionalism and adaptability continued, and contributed to, the fine traditions of Australian servicemen and women.

Afterword

The design, construction and operational use of the Boomerang took place many decades ago. The people, the technology and the geopolitics of that time are far removed from the twenty-first century. Given our remoteness from that era, are there any lessons for the contemporary world from the WWII story of the Boomerang?

In fact, the urgent decision to order the Boomerang highlights the difficulties associated with acquiring appropriate military capabilities to counter threats to national security. This issue is as relevant now as it was in the 1940s. The Australian government ordered the Boomerang

---

1   Weston, *The Australian Aviation Industry*, points out that by 1939, CAC had a world-class aircraft factory, experienced management and staff, and a network of subcontractors. Despite this, in 1940 and 1941, it lacked orders to succeed the Wirraway and Wackett trainer projects. Its manufacturing capacity was again underutilised later in WW II after the Wirraway and Boomerang projects ran down.

at the start of the Pacific War because Australia literally had no modern home defence fighters. The Boomerang was seen as providing "insurance" in case the delivery of fighters from overseas (effectively from the United States) failed. The complete absence of modern fighters was a logical outcome of the policy of focusing the RAAF's home defence mission on countering small-scale maritime raids, a policy that blinded Australia's military establishment to the need to obtain modern fighter aircraft.

The maritime raid policy was disastrously flawed and placed Australia's security at grave risk. Nonetheless, this policy was based on advice given to successive governments by the leaders of Australia's armed services and was supported by senior RAF figures. As this advice was provided from apparently authoritative military sources, it is not surprising the politicians accepted it.

We now know that the maritime raid policy was a disastrous misjudgement despite its seemingly impeccable provenance. This calamity points to two conclusions. Firstly, assessments about a nation's strategic environment, particularly threats to national security, should come from diverse sources and should always be rigorously contested. The views of Cassandras should be welcomed instead of being shut out. Secondly, even the apparently best-qualified threat assessments can simply be wrong. The opaque pseudo-science of threat assessment does not provide mathematically precise predictions.

The Boomerang was hastily ordered in response to the emergence of sudden, unpredictable threats to Australia's national security at the start of the Pacific War in the early 1940s. Events in the 2020s suggest that the fresh emergence of sudden, unpredictable threats to Australia's national security cannot be discounted. Major purchases of military equipment need to be planned and carried out in this cautionary context, where seemingly worst-case scenarios can be replaced with "even worse case scenarios" with very little notice.

APPENDIX: SUMMARY OF BOOMERANG AIR COMBAT ENGAGEMENTS (INCLUDING ATTEMPTED INTERCEPTIONS)

Date Squadron	Pilot(s)	Summary of Engagement
16 May 1943 No. 84 Squadron	Flying Officer Robert Johnstone (A46-51) and Sergeant Maurice Stammer (A46-60)	Johnstone and Stammer were patrolling Merauke, New Guinea. They engaged three Japanese Betty bombers but were unable to damage them.
20/21 May 1943 No. 85 Squadron	Flight Lieutenant Roy Goon (A46-61) and Flying Officer Donald Goode (A46-58)	Attempted night interception: two Boomerangs of No. 85 Squadron's detachment at Potshot airfield, Exmouth Gulf, were scrambled at 2245 in response to radar reports of incoming aircraft. The Boomerangs did not contact the intruders, which were two No. 851 *Kokutai* Emily flying boats. One flying boat dropped 16 x 60-kilogram bombs over Exmouth Gulf while the other evidently saw the Boomerangs as it briefly fired 17 x 20mm rounds in self defence.
21/22 May 1943 No. 85 Squadron	Flying Officer Llewellyn Wettenhall (A46-58) and Flying Officer Malcolm Stevenson (A46-52)	Attempted night interception: two Boomerangs of No. 85 Squadron's Potshot detachment scrambled at 2300 to engage an estimated three enemy aircraft which dropped an estimated nine bombs into the sea about two miles north of Potshot. Pilot Officer Wettenhall reported sighting the exhausts of a Japanese aircraft but could not complete an interception because his fuel ran low. Flying Officer Stevenson did not make any sightings. The intruders were three No. 851 *Kokutai* Emily flying boats, with one of them dropping 16 x 60-kilogram bombs.
30 August 1943 No. 84 Squadron	Possibly Flight Lieutenant Brown (A46-66), Flying Officer Harvey (A46-71), Flight Sergeant Baker (A46-65) and Flight Sergeant Hayes (A46-87)	No. 84 Squadron's ORB briefly states that a Red warning was given at 1010 and an unspecified number of the unit's aircraft scrambled from their Horn Island base and unsuccessfully attempted to intercept the enemy. It also records Brown, Harvey, Baker and Hayes taking off at 1010, suggesting they responded to the Red warning. The cause of the warning is unclear, but a Dinah reconnaissance aircraft had overflown Horn Island the previous day.
6 September 1943 No. 4 Squadron	Flying Officer Thomas Laidlaw (A46-112) and Flying Officer Sydney Carter (A46-?)	On 6 September 1943, Flying Officer Laidlaw and Flying Officer Carter were carrying out a TacR mission near Lae in New Guinea. The two Boomerangs were attacked by Japanese fighters which were likely JAAF Oscars from either the 13th or 24th *Sentai*. Carter escaped the attacking fighters and safely returned to base, but Laidlaw failed to return. His death was confirmed in 1948.
9 September 1943 No. 84 Squadron	Flight Lieutenant Brown (A46-71), Flying Officer Herring (A46-37), Flight Sergeant Johnston (A46-87) and Flight Sergeant Adams (A46-32)	Brown, Herring, Johnston and Adams took off from Merauke airfield at 1035 to join 14 Kittyhawks of No. 86 Squadron to intercept an attack on the airfield by 16 Bettys, which were escorted by 16 fighters. The Kittyhawks reported the destruction of three Japanese fighters, but the Boomerangs were unable to reach the incoming raiders. The bombers attacked the airfield and Boomerang A46-38 was destroyed by bombs.

Date Squadron	Pilot(s)	Summary of Engagement
15/16 September 1943 No. 85 Squadron	Flight Lieutenant Wilson (A46-61), Pilot Officer Denny (A46-77) and Flight Sergeants Felgenhaur (A46-76) and Schoon (A46-70)	Attempted night interception: Wilson, Denny, Felgenhaur and Schoon took off from Potshot at 2345 after a Yellow warning of incoming enemy aircraft at 2330. Red warning at 0015. No sightings of enemy aircraft and All Clear given at 0130. The intruder was a No. 851 *Kokutai* Emily flying boat.
15 November 1943 No. 4 Squadron	Flying Officer Robert Stewart (A46-136)	Stewart was conducting an artillery reconnaissance in the Finschafen area, accompanied by Flying Officer Munro (A46-132) and an escort of two P-40 Kittyhawks. Near Finschhafen airfield he was attacked by a P-38 Lightning of the 9th Fighter Squadron, USAAF. The P-38 was flown by Major Gerald Johnson, a leading ace who ended the Pacific War with a total of 22 (enemy) aircraft destroyed. Johnson, flying a reciprocal course, opened fire and hit the Boomerang, resulting in the port wing burning and the cannon ammunition exploding. Fortunately, Stewart guided the Boomerang to a successful belly landing in scrub. He escaped from the cockpit a few seconds before the aircraft exploded in a fireball, surviving with slight facial injuries.
26 November 1943 No. 4 Squadron	Flight Sergeant Alan Salter (A46-109) and Flying Officer Hector Munro (A46-132)	On 26 November 1943 Salter and Munro were carrying out a TacR mission in the Sanga River area of New Guinea. The Boomerangs were escorted by four P-39 Airacobras of the 41st Fighter Squadron, USAAF, but the escorts were overwhelmed by about twenty JAAF Oscars from the 13th, 59th and 248th *Sentai*. The Airacobra pilots last saw Salter and Munro being pursued by seven Oscars at 100 feet, near the mouth of the Sanga River. No trace of the Boomerangs or pilots was ever found, and they likely crashed into the ocean after being shot down.

SOURCES

NATIONAL ARCHIVES OF AUSTRALIA, VARIOUS:

War Cabinet files

RAAF Operational Record Books

Casualty files

RAAF service records

Aircraft accident files

Aircraft status cards

Various files about the Australian aircraft industry and Boomerang production

Various RAAF reports and files

UNITED KINGDOM NATIONAL ARCHIVES

Aircraft Production: Dominion Requirements

BOOKS

Coulthard-Clark, Chris, and Australia. *Royal Australian Air Force, The third brother : the Royal Australian Air Force 1921-39*, Allen & Unwin in association with the Royal Australian Air Force, North Sydney, 1991.

Dexter, David, *Australia in the War of 1939–1945. Series 1 – Army, Volume VI – the New Guinea Offensives*, Australian War Memorial, Canberra, 1961.

Gill, Hermon, Australia in the War of 1939–1945. Series 2 – Navy, Vol.1 Royal Australian Navy 1939–1942, Australian War Memorial, Canberra, 1957.

Gillison, Douglas, *Australia in the War of 1939–1945, Series 2 – Air, Volume 1, Royal Australian Air Force, 1939-1942*, Australian War Memorial, Canberra, 1962.

Gunston, Bill, *The Illustrated Directory of Fighting Aircraft of World War II*, Salamander Books Ltd, London, 2001.

Hill, Brian, *Wirraway to Hornet: A history of the Commonwealth Aircraft Corporation Pty Ltd 1936 to 1985*, Southern Cross Publications, Bulleen, Victoria, 1998.

Johnston, Mark, *Whispering death*, Allen & Unwin, Crows Nest, NSW, 2011.

Meggs, Keith, *Australian-built aircraft and the industry: Volume 2: Commonwealth Aircraft Corporation. Books 1 and 2*, Echelon Starboard Publications, Nimbin, NSW, 2020.

Mellor, David, *Australia in the War of 1939 – 1945, Series 4 – Civil, Volume 5, The Role of Science and Industry*, Australian War Memorial, Canberra, 1958.

Odgers, George, *Australia in the War of 1939–1945. Series 3 – Air, Volume 2, Air War Against Japan 1943–1945*, Australian War Memorial, Canberra, 1968.

RAAF, *Boomerang Interceptor Overhaul and Repair Manual*, RAAF Publication No.256, Commonwealth Aircraft Corporation Pty. Ltd., Melbourne, June 1943.

RAAF, Manual of Operating Instructions for the Boomerang Interceptor Aircraft, RAAF Publication No. 257, Commonwealth Aircraft Corporation Pty. Ltd., Melbourne, January 1943.

RAAF Historical Section, *Units of the Royal Australian Air Force: a concise history, Volume 2 Fighter Units*, Australian Government Publishing Service Press, Canberra, 1995.

Ross, Andrew, *Armed and Ready. Industrial development and defence of Australia, 1900-1945*, Turton & Armstrong, Wahroonga, NSW, 1995.

Stephens, Alan, *Power plus attitude: ideas, strategy and doctrine in the Royal Australian Air Force 1921-1991*, Australian Government Publishing Service, Canberra, 1992.

Weston, Brian and Australia. *Royal Australian Air Force. Aerospace Centre, The Australian aviation industry: history and achievements guiding defence and aviation policy*, Aerospace Centre, Fairburn, ACT, 2003.

Wackett, Lawrence, *Aircraft Pioneer: An Autobiography – Lawrence James Wackett*, Angus and Robertson, Sydney, 1972.

Williams, Richard, *These are Facts: The Autobiography of Air Marshal Sir Richard Williams*, KBE, CB, DSO, Australian War Memorial and Australian Government Publishing Service, Canberra, 1977.

Wilson, Stewart, *Wirraway, Boomerang and CA-15 in Australian Service*, Aerospace Publications, Weston Creek, ACT, 1991.

ARTICLES

Blainey, Geoffrey and Smith, Ann, Lewis, Essington (1881–1961), *Australian Dictionary of Biography*, National Centre of Biography, Australian National University, 1986.

Kightly, James, *Database: CAC Boomerang*: Aeroplane, Vol 44, No. 8, Issue No. 520, 2016.

Post, Alex, *Wackett, Sir Lawrence James (1896–1982)*, Australian Dictionary of Biography, National Centre of Biography, Australian National University, 2012.

Wingspan International, January 2002.

ONLINE RESOURCES

ADF Serials website; Australian War Memorial website; NLA website; etc.

INDEX OF NAMES